2nd Edition

MANUAL

on the

CAUSES *and* CONTROL

of

ACTIVATED SLUDGE
BULKING *and* FOAMING

David Jenkins
Professor of Environmental Engineering
University of California at Berkeley

Michael G. Richard
RBD Inc., Engineering Consultants
Fort Collins, Colorado

Glen T. Daigger
Vice President and Director
Wastewater Reclamation
CH2M HILL
Denver, Colorado

LEWIS PUBLISHERS
Boca Raton   Ann Arbor   London   Tokyo

Library of Congress Cataloging-in-Publication Data

Jenkins, David, 1935–
    Manual on the causes and control of activated sludge bulking and foaming / by David Jenkins, Michael G. Richard, Glen T. Daigger. — 2nd ed.
        p. cm.
    Includes bibliographical references and index.
    1. Sewage — Purification — Activated sludge process. 2. Sludge bulking. 3. Flocculation.
I. Richard, Michael G. II. Daigger, Glen T. III. Title.
TD756.J46   1993
628.3′54 — dc20   92–40131
ISBN 0-87371-873-9

LEWIS PUBLISHERS, INC.
121 South Main Street, Chelsea, Michigan 48118

PRINTED IN THE UNITED STATES OF AMERICA
    5 6 7 8 9 0

Printed on acid-free paper

To Samantha and Maggie Jenkins,
Joan Richard, and Patty Daigger

David Jenkins is Professor of Environmental Engineering at the University of California at Berkeley, where he has taught since 1960. He came to the United States from England after receiving a PhD in Public Health Engineering from the University of Durham, Kings College and a BSc in Applied Biochemistry from Birmingham University. Dr. Jenkins' research has been in the general areas of biological wastewater treatment, and water and wastewater chemistry. He is the author of over 160 publications. His research work has been recognized by awards from the Water Environment Federation (Eddy, Gasgoine, and Camp medals), the American Society of Civil Engineers (Freeze medal), and the International Association on Water Quality (Sam Jenkins medal). The results of his research, especially those on activated sludge bulking and foaming, have been applied worldwide in the design, upgrading, and operational improvements of biological wastewater treatment plants.

Michael G. Richard is a waste treatment process control specialist with RBD Engineering Consultants in Fort Collins, Colorado. His education includes an AB in field biology and ecology and an MS and PhD in environmental health sciences from the University of California at Berkeley. Prior to his current position, Dr. Richard was a research specialist in wastewater treatment microbiology with David Jenkins at Berkeley and an Assistant Professor of Environmental Health at Colorado State University. He is the author of more than 60 papers, presentations, and technical reports on wastewater treatment microbiology, and was the co-recipient of the Water Environment Federation Eddy medal for research on low F/M bulking. Dr. Richard has been involved in the diagnosis and correction of wastewater treatment microbiology problems for more than 300 cities and 100 industries. He has developed and offered wastewater treatment microbiology training courses at more than 50 locations in the United States, Canada, and South Africa.

Glen T. Daigger is a Vice President with CH2M HILL, an international consulting engineering firm, where he currently serves as Director, Wastewater Reclamation. In this position he is responsible for the nature and quality of the firm's wastewater management services to its municipal clients. He also serves as a process engineer and senior consultant on a wide variety of municipal and industrial wastewater treatment and reclamation projects. Dr. Daigger is an expert in wastewater treatment process and facility design, with special emphasis in biological processes. Areas of special expertise include nutrient control, fixed film systems, and sludge bulking control. He has organized major firm-wide technical initiatives in the areas of nutrient control, advanced techniques for wastewater treatment plant analysis, comprehensive approaches to effluent management, and toxics control in wastewater treatment plants. Dr. Daigger is the author or coauthor of numerous technical articles, several manuals that are widely used in the wastewater treatment profession, and three books. He was educated at Purdue University, where he received his BSCE, MSCE, and PhD in Environmental Engineering. He is a member of the Water Environment Federation, American Society of Civil Engineers, International Association on Water Quality, American Water Works Association, and the Association of Environmental Engineering Professors.

Contents

List of Figures

List of Tables

Preface

The activated sludge process is the most widely used secondary wastewater treatment process in the world. Its proper and efficient functioning relies on the performance of two major treatment units—the aeration basin and the solids separation device—usually a gravity sedimentation basin. Until quite recently most of the research and technical investigation on the activated sludge process had been concerned with factors that affect the efficiency of pollutant removal and the rate of biomass growth in the aeration basin. This focus was maintained even though many investigators and practitioners acknowledged that the great majority of problems of poor activated sludge effluent quality resulted from the inability of the secondary clarifier to efficiently remove the activated sludge biomass from the treated wastewater.

A significant impediment to making advances in understanding the behavior of activated sludge in solids separation processes was the lack of characterization of the complex biocoenosis that makes up the activated sludge biomass. It was, and still is, common practice to regard activated sludge as "suspended solids" or "volatile suspended solids." A key advance in this regard was the work of D.H. Eikelboom at Delft in the Netherlands, who developed a key for characterizing filamentous organisms in activated sludge. This made possible much of the recent work on the resolution of activated sludge solids separation problems. The other key advances in the knowledge had to do with the realization that conditions in the aeration basin (i.e., wastewater characteristics, environmental conditions, aeration basin configuration) influenced the way in which the activated sludge suspended solids behaved during solids separation by gravity settling. Thus the design of an activated sludge system to produce efficient solids separation became one in which there was a proper integration of aeration basin and secondary clarifier designs.

This manual deals with all aspects of activated sludge solids separation problems. It starts with a chapter that describes the nature of activated sludge and its solids separation problems; it traces their origins to fundamental physical, chemical and microbiological properties of the activated sludge. The important role of filamentous organisms is discussed. Next, in Chapter 2, detailed methods for the microscopic characterization of activated sludge for diagnosis of solids separation problems and its "general health" are presented. This section is liberally illustrated with both black and white pictures and color plates. Identification keys are given for filamentous organisms. An extended discussion of the use of filamentous organism characterization in problem diagnosis is presented.

Chapter 3 is a discussion on how to remedy and prevent activated sludge bulking. The theory of activated sludge settling in secondary clarifiers is discussed and is used to develop methods of clarifier and aeration basin management for combatting bulking sludge. The use of polymers, chlorine, and hydrogen peroxide for bulking control is presented, together with many case histories from prototype plants. The resolution of specific bulking problems due to nutrient deficiency, sulfide, low dissolved oxygen concentration, and completely mixed aeration basins is given in detail, again with illustrative case histories. The effect of aeration basin configuration on activated sludge settling and the design and use of selectors for bulking control is covered in depth; illustrative examples from prototype plants are presented.

The final chapter discusses the causes and controls of activated sludge foaming. The role of surfactants, microorganisms such as *Nocardia* spp. and *Microthrix parvicella* is outlined; the effect of physical details of the aeration basin, secondary clarifier, and in-plant recycles on foam retention and foam recycling is discussed. The control of *Nocardia* foaming by using low sludge ages and selector systems of various types is discussed with case histories.

The manual contains an extensive bibliography on all aspects of activated sludge solids separation problems.

This manual is designed to be useful to the wide variety of professionals who design, manage, monitor, and operate the activated sludge process. There is material in this text that is useful for laboratory personnel, for design engineers, and for treatment plant operators. The manual serves as a useful text on activated sludge microbiology and treatment plant design and operation. It has been used widely and successfully in laboratory short courses on activated sludge characterization.

Acknowledgments

This manual was made possible by the work and support of many individuals and organizations. Our sincere thanks and gratitude are due to them. The first edition of the manual was made possible by grants from the Water Research Commission of the Republic of South Africa and the Environmental Protection Agency of the United States of America.

Much of the research that was the foundation for this manual was conducted over the last two decades by a dedicated and imaginative group of graduate students and postdoctoral researchers at the University of California at Berkeley. Although their names appear in the Bibliography they, of all people, deserve individual mention. They are Mesut Sezgin, Denny Parker, John Palm, Tony Lau, Peter Strom, Sang-Eun Lee, Ben Koopman, Oliver Hao, Michael Richard, J.B. Neethling, Greg Shimizu, Kay Johnson, Andre van Niekerk, Bernardo Vega-Rodriguez, Haro Bode, Y-J. Shao, Paul Pitt, Daniel Cha, Daniel Mamais, Linda Blackall, Valter Tandoi, Mark Hernandez, Cho-Fei Ho, Jean Weber, and Krishna Pagilla.

The application of new ideas into wastewater treatment practice often requires courage and a degree of blind faith on the part of the treatment plant personnel. Thanks are due to the many people who have willingly tried out the results of our research in practice. Special recognition is due to the people who took special risks in "doing it first." These are Russ Edwards, City of Albany, Georgia; Bob Beebe, San Jose/Santa Clara, California; Randy Gray, Stroh Brewing Co., Longview, Texas; Bill Keaney, N. San Mateo Sanitation District, California; Mike Wheeler, City of Hamilton, Ohio; Millard H. Robbins, Upper Occoquan Sewage Authority, Centreville, Virginia; Michael Read and Richard Stillwell, Clackamas Service District, Oregon; Gary Vaughn, Fayetteville, Arkansas; Dale Richwine and Carlo Spani, United Sewerage Agency, Oregon; John Reid, Stone Container Corporation, Ontonagon, Michigan; Todd Cockburn, City of San Francisco, California; and Wendell Kido, City of Sacramento, California.

We would also like to acknowledge the contributions of our peers who freely and openly discussed our ideas and often provided us with fresh insights. These include Orrie Albertson, Denny Parker, Tom Wilson, Wes Eckenfelder, David Stensel, and Bob Okey.

The microphotographs were taken by Michael Richard. CH2M HILL produced the figures. Sarah Jenkins Muren and Joan Jenkins prepared the manuscript. Joan Jenkins verified the citations. The patience, skill and hard work of all these people is acknowledged.

David Jenkins, Berkeley, California
Michael Richard, Fort Collins, Colorado
Glen Daigger, Denver, Colorado

Glossary

BOD	. . . .	Biochemical Oxygen Demand
COD	. . . .	Chemical Oxygen Demand
DO	. . . .	Dissolved Oxygen
DSVI	. . . .	Diluted Sludge Volume Index
EBPR	. . . .	Enhanced Biological Phosphorus Removal
F/M	. . . .	Food to Microorganism Ratio
FTU	. . . .	Formazine Turbidity Units
g	. . . .	Gram
hr	. . . .	Hour
L	. . . .	Liter
m	. . . .	Meter
M	. . . .	Molar
MCRT	. . . .	Mean Cell Residence Time
mg	. . . .	Milligram
MG	. . . .	Million Gallons
MGD	. . . .	Million Gallons per Day
min	. . . .	Minutes
mL	. . . .	Milliliter
MLSS	. . . .	Mixed Liquor Suspended Solids
mm	. . . .	Millimeter
mM	. . . .	Millimolar
μ	. . . .	Micromolar
μg	. . . .	Microgram
μm	. . . .	Micron
PHB	. . . .	Poly-β-hydroxybutyric Acid
PVC	. . . .	Polyvinylchloride
RAS	. . . .	Return Activated Sludge
sec	. . . .	Second
sp., spp	. . . .	Specie, Species
SS	. . . .	Suspended Solids
SSVI	. . . .	Specific Sludge Volume Index
SV30	. . . .	Settled Volume at 30 min
SVI	. . . .	Sludge Volume Index
TEFL	. . . .	Total Extended Filament Length
TKN	. . . .	Total Kjeldahl Nitrogen
TOC	. . . .	Total Organic Carbon
VSS	. . . .	Volatile Suspended Solids
WAS	. . . .	Waste Activated Sludge
$SSVI_{3.5}$	. . . .	Sludge Volume Index at 3.5 gSS/L
SVI_m	. . . .	Mallory Sludge Volume Index

2nd Edition

MANUAL
on the
CAUSES *and* CONTROL
of
ACTIVATED SLUDGE
BULKING *and* FOAMING

". . . there is nothing I can do about it now!"
Willie Nelson

CHAPTER 1

Solids Separation Problems

INTRODUCTION

The activated sludge process always consists of two liquid stream processing units—the aeration basin (biological reactor) and the secondary clarifier. Often there is also a primary clarifier in the flow sheet. The aeration basin provides an environment for the removal and transformation of both soluble and particulate pollutants by a mixed and variable consortium of micro- and macro-organisms called activated sludge. The secondary clarifier provides a quiescent environment which allows the activated sludge solids to separate by flocculation and gravity sedimentation from the treated wastewater. The objective of this basin is to provide a clarified [low suspended solids (SS), low turbidity] overflow called the secondary effluent and a thickened underflow called the return activated sludge (RAS). For efficient treatment both the aeration basin and the secondary clarifier must function satisfactorily. Therefore, factors affecting both biological oxidation and solids separation are important in determining overall process efficiency.

It is the purpose of this manual to describe the nature of activated sludge solids separation problems, to discuss their causes and to present a rational approach for preventing these problems by proper design and operation, or for solving them when they occur. The manual is illustrated liberally with case histories.

SOLIDS SEPARATION PROBLEMS

There are several types of activated sludge solids separation problems. These problems have been named in terms of the effects that they cause in the activated sludge treatment process and because of this, their definition is not rigorous. The physical/microbiological causes for each of these problems have been fairly well defined. Table 1 presents a summary of the causes and effects for the most commonly identified activated sludge solids separation problems.

THE ACTIVATED SLUDGE FLOC

Most activated sludge solids separation problems can be related to the nature of the activated sludge floc. In typical activated sludge there is a wide range of particle sizes—all the way from single bacteria with dimensions in the approximate range of 0.5 to 5 μm up to large aggregates (flocs) that can reach sizes of more than 1 mm (1000 μm).

Activated sludge flocs are made up of two types of components: a biological component consisting of a wide variety of bacteria, fungi, protozoa and some metazoa and a nonbiological component made up of inorganic and organic particulates. The basis of the floc appears to be a number of heterotrophic bacteria that include genera such as *Pseudomonas*, *Achromobacter*, *Flavobacterium*, *Alcaligenes*, *Arthrobacter*, *Citromonas*, and *Zoogloea* (Dias and Bhat, 1964; Pike, 1972; Tabor, 1976). Early suggestions that a single floc-forming organism, *Zoogloea ramigera*, was the sole basis of the activated sludge floc have been discounted, although zoogloeas often are observed in activated sludge (Williams and Unz, 1983) and are one sign that the so-called "selector-effect" exists (van Niekerk et al., 1987) or that the organic loading is high. At typical rates of operation, microorganism viability in the floc is low, in the range of 5% to 20% (Weddle and Jenkins, 1971). In the absence of chemical addition in municipal wastewater treatment systems, activated sludge volatile matter contents range from approximately 60% to 90%. Besides microorganisms, activated sludge flocs contain organic and inorganic particles from the incoming wastewater and extracellular "polymers" that, it has been suggested, play a role in the bioflocculation of the activated sludge (Pavoni et al., 1972; Unz and Farrah, 1976; Farrah and Unz, 1976; Tago and Aiba, 1977). These extracellular polymers in the flocs typically are composed mostly of carbohydrates and contribute 15% to 20% by weight of the MLSS.

On the basis of visual observation and some physical measurements it has been suggested that

Table 1. Causes and Effects of Activated Sludge Separation Problems

Name of Problem	Cause of Problem	Effect of Problem
Dispersed growth	Microorganisms do not form flocs but are dispersed, forming only small clumps or single cells.	Turbid effluent. No zone settling of the activated sludge.
Slime (jelly) viscous bulking; (also has been referred to as nonfilamentous bulking)	Microorganisms are present in large amounts of exocellular slime. In severe cases the slime imparts a jelly-like consistency to the activated sludge.	Reduced settling and compaction rates. Virtually no solids separation in severe cases, resulting in overflow of sludge blanket from secondary clarifier. Sometimes a viscous foam is also present.
Pin floc or pinpoint floc	Small, compact, weak, roughly spherical flocs are formed, the larger of which settle rapidly. The smaller flocs settle slowly.	Low sludge volume index (SVI) and a cloudy, turbid effluent.
Filamentous bulking	Filamentous organisms extend from flocs into the bulk solution and interfere with compaction, settling, thickening and concentration of activated sludge.	High SVI—very clear supernatant. Low RAS and WAS solids concentrations. In severe cases the sludge blanket overflows the secondary clarifier. Solids handling processes become hydraulically overloaded.
Blanket rising	Denitrification in secondary clarifier releases poorly soluble N_2 gas which attaches to activated sludge flocs and floats them to the secondary clarifier surface.	A scum of activated sludge forms on surface of the secondary clarifier.
Foaming/scum formation	Caused by (i) nondegradable surfactants and by (ii) the presence of *Nocardia* spp., *Microthrix parvicella*, or type 1863.	Foams float large amounts of activated sludge solids to surface of treatment units. *Nocardia* spp. and *M. parvicella* foams are persistent and difficult to break mechanically. Foams accumulate and can putrefy. Solids can overflow into secondary effluent or overflow tank freeboard onto walkways.

there are two levels of structure in the activated sludge floc. These have been termed the "microstructure" and the "macrostructure" (Sezgin et al., 1978). The microstructure is imparted by processes of microbial adhesion, aggregation, and bioflocculation. It is the basis for floc formation because, without the ability of one microorganism to stick to another, large aggregates of microorganisms such as exist in activated sludge would never form. While the mechanism of bioflocculation in activated sludge is complex and the factors affecting it are poorly understood, it is likely that it results from the interaction (bridging) between extracellular microbial

polymers functioning as polyelectrolytes. These extracellular microbial polymers have been termed "the glycolax" by Costerton and Irvin (1981) and they form a felt-like envelope around individual cells or groups of cells. The polymers are various types of polysaccharide and glycoprotein fibers.

For an activated sludge in which only "floc-forming" bacteria are present, i.e., flocs that have only the so-called microstructure, the flocs are usually small (up to about 75 μm in dimension), spherical and compact (Figures 1a and 1b). Under some circumstances, especially when the bioflocculation is not well developed, they can be

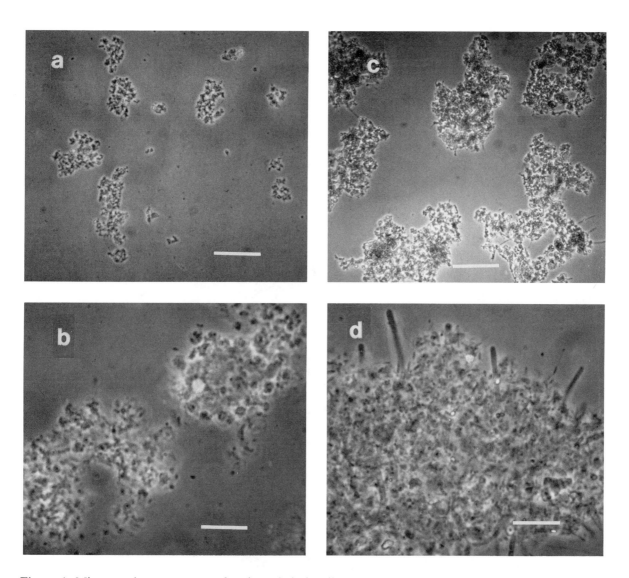

Figure 1. Microscopic appearance of activated sludge flocs: *a.* small, weak flocs (pin floc) (100X); *b.* small, weak flocs (1000X); *c.* flocs containing filamentous organisms (100X); and *d.* flocs containing filamentous organism "network" or "backbone" (1000X phase contrast; bar = 100 μm).

sheared in the turbulent environment of an activated sludge aeration basin. In high mean cell residence time (MCRT) activated sludge systems, this type of floc gives rise to the condition termed "pin floc" or "pinpoint floc." As indicated in Table 1, this type of sludge settles rapidly but can produce a turbid supernatant. It is the larger compact flocs that settle rapidly; the smaller aggregates, sheared off from larger flocs, settle slowly and create the turbid supernatant.

Sezgin et al. (1978) proposed that the macrostructure of activated sludge flocs is provided by filamentous microorganisms. These organisms form a network or backbone within the floc onto

which the floc-forming bacteria cling, just like flesh on a bone. When an activated sludge culture contains filamentous organisms, large floc sizes are possible because the filamentous organism backbone provides the floc with strength so that it can hold together in the turbulent environment of the aeration basin. Large flocs containing filamentous organisms also become irregularly shaped (Figure 1c) rather than approximately spherical (as when filamentous organisms are absent or present only in small numbers). The filamentous organism network influences the shape of the floc because the floc grows in the same direction that the filamentous organism net-

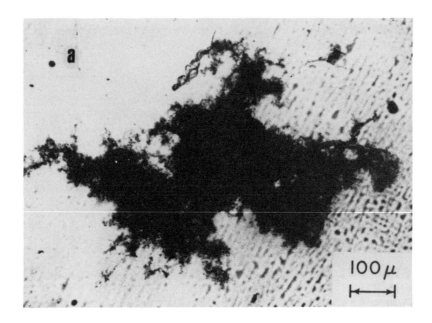

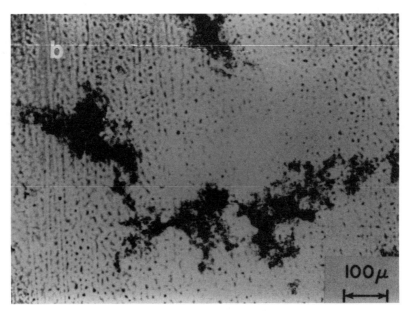

Figure 2. Effect of shear rate on 1400 μm-size activated sludge flocs: *a*. G = 10.5 sec^{-1}; *b*. G = 79 sec^{-1}.

work grows. For example, note the irregular filament-directed growth of the floc shown in Figure 1c. Sezgin (1977) showed that the shape of activated sludge flocs changed from approximately spherical at low filamentous organism levels to approximately cylindrical at high filamentous organism levels, and proposed that the cylindrical shape was due to the floc growth being directed by filaments which themselves are cylindrical. Parker et al. (1971) demonstrated the effect of filamentous organism structure in pre-

serving the integrity of activated sludge flocs under conditions of increasing shear (Figure 2). In Figure 2 it can be seen that as the mean velocity gradient (a) is increased, the floc-forming bacteria are stripped away from their filamentous organism backbone.

Support for the effect of filamentous organisms on the structure and shape of activated sludge flocs was provided by Lau et al.(1984a), who grew a floc-forming bacterium isolated from activated sludge (*Citromonas* spp.) in dual axenic

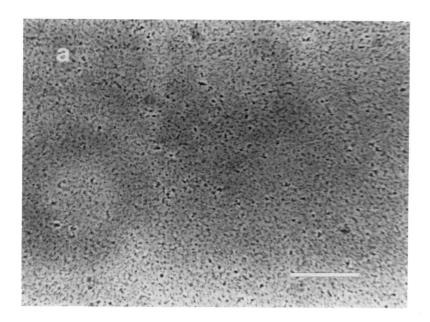

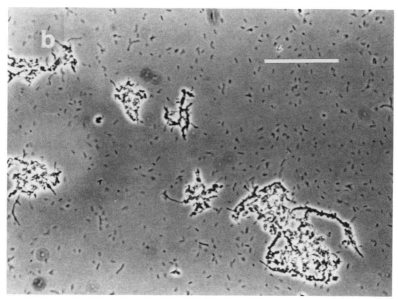

Figure 3. Effect of filamentous organisms on floc structure:
a. aggregates observed for a pure culture of an activated sludge floc former; *b*. aggregates observed for a dual culture of a single floc former and a single filamentous organism (100X phase contrast; bar = 100 μm) (Lau et al., 1984).

culture with a *Sphaerotilus natans* strain also isolated from activated sludge. When either organism was grown singly in chemostat cultures the aggregates formed were compact and roughly spherical (Figure 3a). When grown together in approximately balanced growth, irregularly-shaped flocs reminiscent of those observed in activated sludge were formed (Figure 3b). Filamentous backbones were observed at the center of these irregularly shaped flocs.

The activated sludge solids separation problems described in Table 1 can be interpreted in terms of failure of either the microstructure or the macrostructure of the activated sludge floc.

Dispersed growth is caused by a microstructure failure in which, for some reason, microorganisms do not stick to each other. Bioflocculation does not occur. Often this is due to the selection of nonflocculating bacteria, either single cells or filaments, under specific activated sludge growth conditions. Poorly biodegradable surfactants can cause dispersed growth by dispersing (deflocculating) already flocculated activated sludge.

Slime or jelly formation has been termed "viscous bulking" by Hale and Garver (1983) and is recognized by Eikelboom and van Buijsen (1981) as "zoogloeal bulking." It is caused by a microstructure failure in which too much of the extracellular material that contributes to bioflocculation is produced. Microbial cells are "dispersed" in an extracellular mass which can become highly water retentive. Since this condition can result in a viscous, poorly-settling and compacting activated sludge, it is possible that it constitutes one of the conditions that previously has been referred to as nonfilamentous bulking (Pipes, 1979). This condition can be visualized microscopically by reverse staining with India ink. Figure 4c shows an unstained wet mount of an extremely viscous activated sludge viewed with a light microscope at a magnification of 100X using phase contrast illumination. Notice that the flocs contain masses of dispersed cells and also have smooth rounded margins in comparison with the "normal" activated sludge floc (Figure 4a).

When a drop of India ink is added to a live preparation of a "normal" activated sludge floc, the carbon black particles in the ink penetrate deeply into the floc, obscuring discernible structures (Figure 4b). When there is a large amount of extracellular material present the carbon black particles do not penetrate the floc (Figure 4d and 4e). In Figure 4e the individual rod-shaped bacteria can be seen, surrounded by large masses of extracellular material. The presence of an excessive amount of extracellular material can be confirmed by analyzing the activated sludge for polysaccharide using the anthrone test as described in Table 13 (Chapter 2).

Normal activated sludges treating domestic wastewater contain approximately 15% to 20% carbohydrate as glucose on a VSS weight basis. Activated sludges suffering from viscous bulking have carbohydrate levels in the range from 25% to 60% of the VSS.

Pin floc or pinpoint floc is due to a macrostructure failure that usually occurs at very low organic loading but may, in some industrial wastewater activated sludge systems, be caused by chemical interference in floc formation (i.e., dispersion). Basically, there is no macrostructure because there are no (or very few) filamentous organisms present. The floc relies solely on the microstructure (glycolax) to maintain its integrity and, if this is lacking, the flocs can break up easily. The flocs are small, roughly spherical, weak and readily broken up and sheared in the aeration basin (Figures 1a and 1b).

Bulking is a macrostructure failure in which, in a sense, there is too much macrostructure!! Filamentous organisms that provide the macrostructure are present in large numbers. They interfere with the compaction and settling of the activated sludge either by producing a very diffuse floc structure ("stretched out" or "diffuse floc") or by growing in profusion beyond the confines of the activated sludge floc into the bulk medium and bridging between flocs. Figures 5a and 5b illustrate a bridged floc. Figures 5c and 5d show a "stretched out," diffuse floc. The type of compaction and settling interference depends on the causative filamentous organism.

Table 2 indicates the type of settling interference caused by various filamentous organisms observed in activated sludge. Note that some filamentous organisms can cause both types of bulking.

FOAMING/SCUM FORMATION

Filamentous organism-related foaming problems in activated sludge are associated largely with the presence of *Nocardia* spp. and *M. parvicella*; to a much lesser degree microbially-related foam occurs with type 1863 at very low MCRT [high organic loading (F/M)] usually in the presence of significant oil and grease levels.

Both *Nocardia* spp. and *M. parvicella* possess poorly wettable cell surfaces (hydrophobic surfaces). When they grow in sufficient numbers in activated sludge they render the flocs hydrophobic and amenable to attachment on air bubbles. The air bubble-floc aggregate is less dense than water and therefore floats to the surface. Since it is hydrophobic, once at the surface it tends to stay there; it accumulates to form a thick, chocolate-brown colored float or scum. A similar looking float can be produced by denitrification in a secondary clarifier. Here nitrate (NO_3), produced in the aeration basin, serves as an "oxygen source" for facultative activated sludge bacteria in the sludge blanket at the bottom of the second-

ary clarifier. The NO_3 is converted to the sparingly water-soluble nitrogen gas (N_2) which comes out of solution. It acts as a superb floc flotation device because the N_2 bubbles are very small and

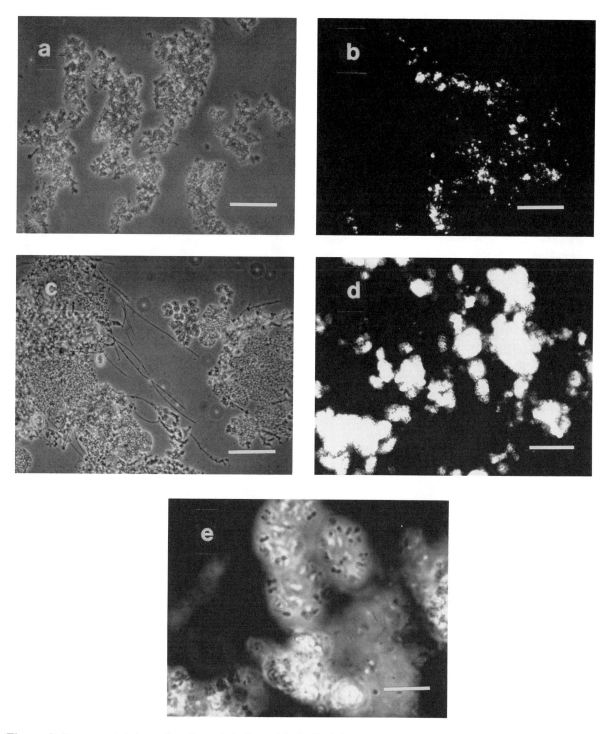

Figure 4. Reverse-staining of activated sludge with India ink: *a.* appearance of "normal" activated sludge, and *b.* the same sludge with India ink; *c.* appearance of "viscous" activated sludge with India ink (*a.−d.* 100X phase contrast; bar = 100 μm; *e.* 1000X phase contrast; bar = 10 μm).

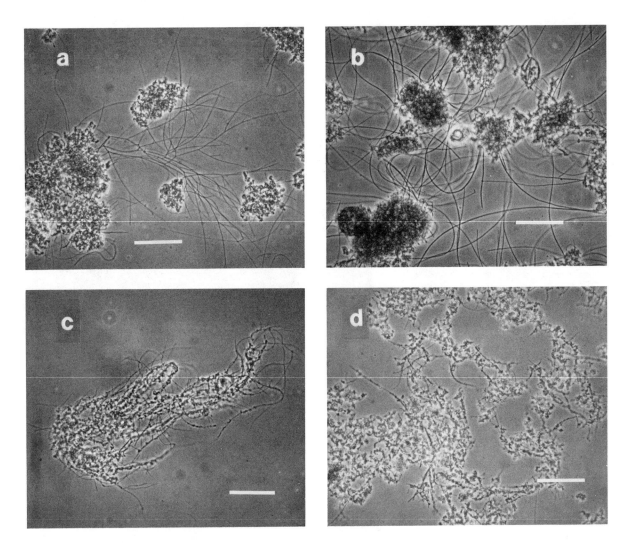

Figure 5. Effect of filamentous organisms in activated sludge on floc morphology and settleability: *a*. and *b*. inter-floc bridging; *c*. and *d*. diffuse floc structure (all 100X phase contrast; bar = 100 μm).

they are produced inside the activated sludge flocs so that they are quite well bound to them. Again, this gas-solid aggregate is lighter than water and so it rises to the surface as a float. Floating sludge problems can be much more severe when the denitrifying sludge also is filamentous. Denitrification scum and *Nocardia*- or *M. parvicella*-derived scum can be differentiated as follows:

Nocardia spp. or M. parvicella scum is associated with:

- large, strong bubbles in the aeration basin
- a marked concentration of the organisms in the scum over their levels in the mixed liquor.

Denitrification scum is associated with:

- small nitrogen gas bubbles in the secondary clarifier
- no difference in filamentous organism abundance between the scum and the mixed liquor.

The type 1863 foam is white-grey and easily collapsed. It occurs on aeration basins and in secondary effluents when a plant is being operated at just about the upper limit of F/M (before soluble BOD breakthrough). Typically, this occurs at MCRTs of below 2 days.

The scum-accompanying nutrient deficiency is sticky and contains a high SS (activated sludge) content. It is likely caused by extracellular poly-

Table 2. Type of Compaction and Settling Interference Caused by Various Filamentous Organisms

Bridging	Open Floc Structure
type 1701	type 1701
type 0041	type 0041
Microthrix parvicella	*Microthrix parvicella*
type 1851	type 1851
Nostocoida limicola	*Nostocoida limicola*
type 021N	type 0675
Sphaerotilus natans	type 0092
type 0961	
type 0803	
Thiothrix spp.	
Haliscomenobacter hydrossis	

mers produced as shunt products by the activated sludge bacteria under nutrient-deficient or high F/M and high DO uptake rate conditions.

TYPE OF FLOC AS AFFECTED BY FILAMENTOUS ORGANISM LEVEL

In simple terms we can conceive of there being three types of activated sludge floc, based on the amount of filamentous organisms that they contain (Sezgin et al. 1978).

Figure 6a represents the *ideal* floc. In it the filamentous organisms and floc-forming organisms grow in "balance." The filamentous organisms grow largely inside the floc, providing it with a structure and with strength. A few filamentous organisms protrude out from the floc surface into the bulk solution but they are not present in sufficient quantities to interfere with the compaction and settling rates of the activated sludge. Such a sludge will have a sludge volume index (SVI) in the region of 80 to 120 mL/g; it will produce a low turbidity and low SS supernatant when it settles. This type of sludge will give a 60-min settling curve with two distinct regions: an initial rapid settling period with a clear sludge interface, followed by a slower settling period when hindered settling and sludge compaction are occurring.

Figure 6b is a pinpoint floc. Here there are few or no filamentous organisms present. The structure is "microstructure" only. The flocs are small and if the glycolax is not properly developed, they are weak and can be readily sheared and broken up in the turbulence of an aeration basin.

The large flocs settle and compact very rapidly; the smaller aggregates do not settle well. Such a sludge has a very low SVI (< 70 mL/g). In a settling test the supernatant usually is turbid and sometimes has a high SS level; the sludge-supernatant interface often will not be distinct.

Note that the presence of extracellular bacterial polymers can result in strong, though small, flocs even in the absence of filaments. This type of floc is often seen in "selector"-activated sludge systems. The terms pin floc or pinpoint floc are reserved for the small weak flocs, without filaments usually seen in high MCRT (extended aeration) systems.

Figure 6c is a filamentous bulking sludge. In it, filamentous organisms grow in profusion. They grow both inside the flocs and outside the flocs, penetrating into the bulk solution. They can stretch out the floc, making it diffuse, and/or bridge in between flocs, thereby interfering with their close approach, just like mechanical protrusions. Such a sludge settles and compacts poorly—it will have a high SVI (> 150 mL/g). When such a sludge settles (at all) it produces an extremely clear supernatant (what there is of it), because the large numbers of extended filamentous organisms filter out the small particles that cause turbidity.

MEASUREMENT OF FILAMENTOUS ORGANISM LEVELS IN ACTIVATED SLUDGE

It has been stated above that the overall filamentous organism level in activated sludge is extremely important in determining its settling and compacting characteristics. Several workers have provided experimental evidence in support of this.

Finstein and Heukelekian (1967) were the first investigators to determine a quantitative relationship between filamentous organism level and activated sludge-settling properties. They showed that the SVI of activated sludge could be related to the total filament length per floc.

Pipes (1979) microscopically counted the numbers of filamentous organisms extending from activated sludge flocs using a Neubauer haemocytometer, and found that at low filament numbers (10^2 filaments/mg VSS) the SVI was below 100 mL/g, while at high filament numbers ($> 10^2$–10^3 filaments/mg VSS) the SVI increased markedly.

Sezgin et al. (1978), Palm et al. (1980) and later

A. IDEAL NONBULKING ACTIVATED SLUDGE FLOC

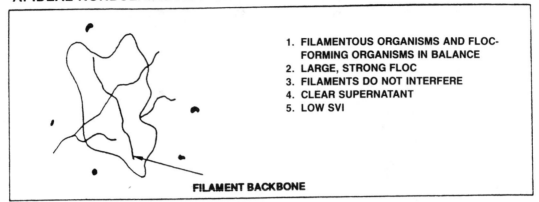

1. FILAMENTOUS ORGANISMS AND FLOC-FORMING ORGANISMS IN BALANCE
2. LARGE, STRONG FLOC
3. FILAMENTS DO NOT INTERFERE
4. CLEAR SUPERNATANT
5. LOW SVI

FILAMENT BACKBONE

B. PIN POINT FLOC

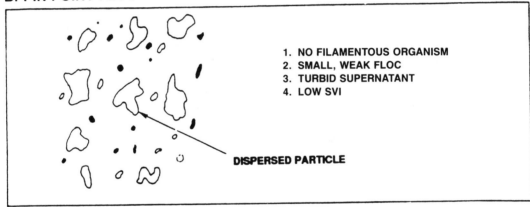

1. NO FILAMENTOUS ORGANISM
2. SMALL, WEAK FLOC
3. TURBID SUPERNATANT
4. LOW SVI

DISPERSED PARTICLE

C. FILAMENTOUS BULKING ACTIVATED SLUDGE

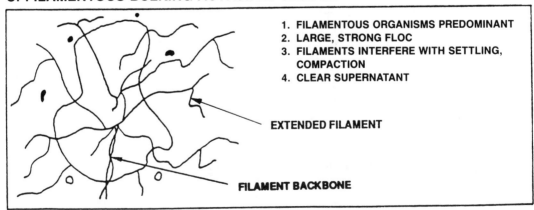

1. FILAMENTOUS ORGANISMS PREDOMINANT
2. LARGE, STRONG FLOC
3. FILAMENTS INTERFERE WITH SETTLING, COMPACTION
4. CLEAR SUPERNATANT

EXTENDED FILAMENT

FILAMENT BACKBONE

Figure 6. Effect of filamentous organisms on activated sludge floc structure.

Table 3. Filament Measurement Technique of Sezgin et al. (1978)

1. Transfer 2 mL of a well-mixed activated sludge sample of known SS concentration using a wide-mouth pipette (0.8 mm diameter tip) to 1 liter of distilled water in a 1.5 liter beaker and stir at 95 rpm on a jar test apparatus (G = 85 sec^{-1}) for 1 min.

2. Using the same pipette, transfer 1.0 mL diluted sample to a microscopic counting chamber calibrated to contain 1.0 mL and cover with a glass cover slip.

3. Using a binocular microscope at 100X magnification with an ocular micrometer scale, count the number of filaments present in the whole chamber or a known portion of it and place these in the following size classifications: 0 to 10 μm, 10 to 25 μm, 25 to 50 μm, 50 to 100 μm, 100 to 200 μm, 200 to 400 μm, 400 to 800 μm, and greater than 800 μm. Measure filaments of greater than 800 μm in length individually.

4. Express results as the total μm of filament length per g MLSS or per mL MLSS:

$$\text{total extended filament length, } \mu\text{m/g MLSS} = \frac{\text{total filament length, } \mu\text{m in the 1.0 mL diluted sample} \times \text{the dilution factor (X500 in the above sample)}}{\text{MLSS concentration, g/L}}$$

$$\text{total extended filament length, } \mu\text{m/mL MLSS} = \text{total filament length, } \mu\text{m, in the 1.0 mL diluted sample} \times \text{dilution factor (X500 in the above example)}$$

Lee et al. (1982) all used the Sezgin et al. method for measuring the total extended filament length (TEFL) in activated sludge. The method is presented in Table 3. All of these investigators found correlations between various measures of activated sludge settling (such as SVI and zone settling velocity) and TEFL for activated sludge grown in the laboratory on a domestic sewage. In general, the SVI increased rapidly above 100 mL/g when TEFL values increased above 10^7 μm/mL (Figure 7). Sezgin et al. (1980) found that these relationships were valid for activated sludge taken from several full-scale plants (Figure 8).

Lee et al. (1983) correlated TEFL with sludge settleability measured by several techniques—the SVI, the diluted SVI (DSVI) (Stobbe, 1964) (Table 4), the SVI at SS concentrations of 1.5, 2.5, and 3.5 g/L and the stirred SVI at 2.5 gSS/L [similar to the SSVI of White (1976)]. Lee et al. reasoned that because TEFL previously had been shown to be a quantitative index of sludge settleability, the settling test to which it best correlated would be the best test for judging sludge settleability. Their data showed that the closest and most consistent correlation was obtained between TEFL and the DSVI. On this basis they proposed the replacement of the standard SVI with the DSVI. Similar conclusions were reached by Matsui and Yamamoto (1983), who used a video imaging system to aid the microscopic counting

of filament length. Further justification for this proposal was provided by Koopman and Cadee (1983), Pitman (1984), and Rachwal et al. (1982) who showed that the parameters relating activated sludge settling velocity and activated sludge SS concentration were well correlated with the DSVI. Thus, a simple test—the DSVI—is available for the prediction of activated sludge separation properties in process units that is well correlated to the level of filamentous organisms present in the activated sludge.

The San Jose/Santa Clara, California, Water Pollution Control Plant (SJ/SCWPCP) staff have simplified filament counting in activated sludge using the method described in Table 5. Basically this method does not attempt to measure total extended filament length. Rather, the number of times that filaments intersect a single line drawn on the microscope eyepiece is counted for one traverse of consecutive fields of view across a wet mount preparation of the activated sludge sample. This technique has been used with good results by the staff of the SJ/SCWPCP. Their data relating "Filament Count" to SVI illustrates two important points. The SVI and "Filament Count" are temporally correlated (Figure 9), thus "Filament Count" cannot be used as an early warning for an increase in SVI. The SJ/SCWPCP consists of two activated sludge plants in series. The first (secondary)

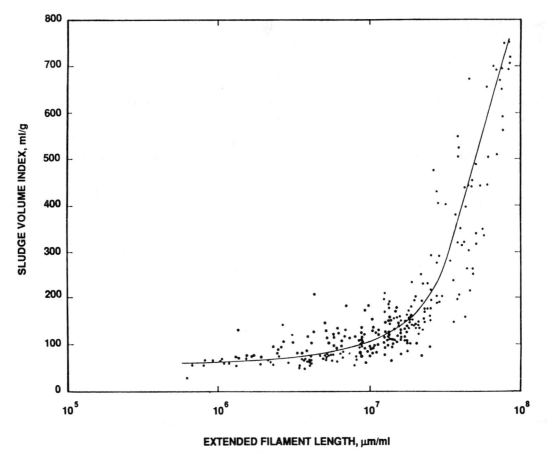

Figure 7. Effect of extended filament length on SVI (Palm et al., 1980). Reprinted with permission of the Water Environment Federation.

plant is for BOD removal; the second (tertiary) plant is designed for complete nitrification of the secondary effluent. The filamentous organisms that occur in these two treatment plants are different and, as Figure 10 shows, there is a different relationship between SVI and "Filament Count." This suggests that the growth form of the filamentous organism in activated sludge or the presence of different filamentous organisms may influence the degree to which a given length of filamentous organisms influences sludge settleability.

Taken together, the observations that TEFL and DSVI (Lee et al. 1983) are well correlated and that increasing filament counts do not precede settleability decreases mean that *filament counting per se of activated sludge* is not an efficient process control procedure. It is far easier to run a DSVI than to count filaments and, since they both give the same information at the same time, it makes sense to do the easier test! On the other hand, filament identification is an extremely useful tool for diagnosing and rectifying activated

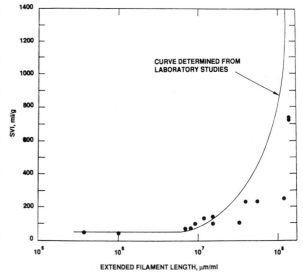

Figure 8. Variation of SVI with filament length at 14 prototype California wastewater treatment plants. Reprinted with permission from Sezgin et al., "The Role of Filamentous Microorganisms in Activated Sludge Settling," *Prog. Water Technol.*, 12, 1980. Pergamon Press plc.

Table 4. Diluted SVI (DSVI) Procedure (Stobbe, 1964)

1. Set up several 1-liter graduate cylinders (the number will depend on prior knowledge of the settleability of the sludge).

2. Using well-clarified secondary effluent, prepare a series of twofold dilutions of the activated sludge (i.e., no dilution, 1:1 dilution, 1:3 dilution).

3. Stir the graduate cylinders individually for 30–60 sec using a plunger to resuspend and uniformly distribute the sludge solids.

4. Allow the activated sludge to settle for 30 min under quiescent conditions.

5. Observe the settled sludge volume (SV_{30}) in the graduate cylinder where the settled volume is less than and closest to 200 mL ($SV_{30} < 200$ mL).

6. Calculate the diluted SVI using:

$$\text{DSVI (mL/g)} = \frac{SV_{30} \text{ (mL/L)} \times 2^n}{\text{SS (g/L)}}$$

where n is the number of twofold dilutions required to obtain a settled sludge volume (SV_{30}) less than 200 mL and SS is the suspended solids concentration of the *undiluted* activated sludge.

Table 5. Simplified Filament Counting Technique Used by the San Jose/Santa Clara Water Pollution Control Plant

1. Transfer 50 μL mixed liquor sample to a glass slide.

2. Cover completely with a 22 x 30 mm cover slip.

3. Using 100X total magnification and starting at the edge of the cover slip, observe consecutive fields across the entire 30 mm length of the cover slip. (At SJ/SCWPCP this is 17 fields.)

4. The eyepiece is fitted with a single hairline. Count the number of times that any filamentous organism intersects with the hairline.

5. Sum the number of intersections for all fields examined. This is the Filament Count. If a "unit count" is desired, the Filament Count must be multiplied by the number of fields in the 22-mm width of the slide. At San Jose this is 12 fields. Thus:

$$\text{Filament Count}/\mu\text{L} = \frac{\text{filament intersections in field counted}}{50 \ \mu\text{L}} \times 12$$

sludge solids separation problems (as described in Chapter 2), and is essential for differentiating filamentous bulking from polysaccharide (slime) bulking.

NOCARDIA FILAMENT COUNTING

The rapid assessment of *Nocardia* populations in activated sludge can be made by a *Nocardia* filament counting technique developed by Vega-Rodriguez (1983) and Pitt and Jenkins (1990) (Table 6). Other methods for determining *Nocardia* levels in activated sludge are either laborious, slow, and not sensitive enough (e.g., plate counting on selective media and counting colonies; Wheeler and Rule, 1980) or inappropriate (e.g., foaming tests that are influenced by physical/chemical factors as well as by *Nocardia* levels and also are highly sensitive to equipment design and operation). Pitt and Jenkins' technique basically consists of microscopic counting of the number of branched Gram-positive filaments of length

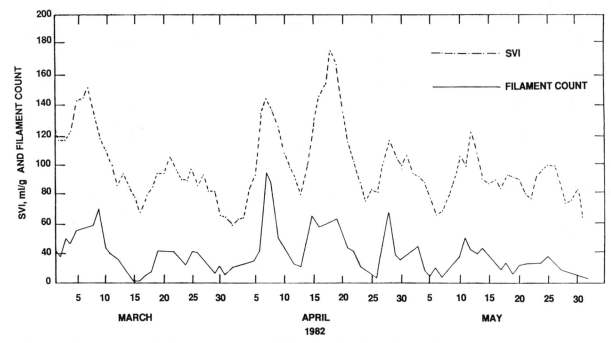

Figure 9. Comparison of filament count and SVI at the San Jose/Santa Clara, CA water pollution control plant. (Beebe et al., 1982).

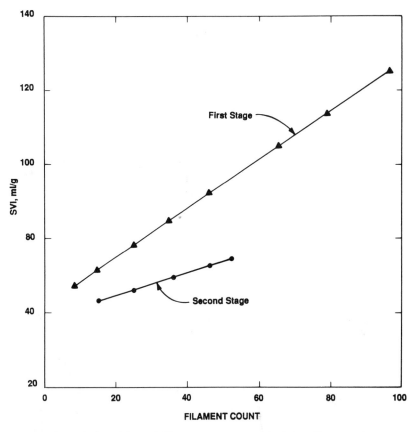

Figure 10. Relationships of filament count to SVI for first- and second-stage activated sludge plants, San Jose/Santa Clara, CA water pollution control plant.

Table 6. *Nocardia* **Filament Counting Technique (Pitt and Jenkins, 1990)**

1. Blend 400 mL of mixed liquor of known VSS concentration for 2 to 3 min at low power in a blender.

2. Prepare several (5 to 10) clean frosted microscope slides by marking the edges with a glass scribe at three equally spaced points along their length.

3. On each slide place 80 μL blended mixed liquor using a micropipette. Spread the liquid evenly over the entire nonfrosted area of the slide.

4. Air dry the slides.

5. Microscopically examine the slides at a magnification of 100X using phase contrast to check for even solids distribution over the slide. Discard slides showing uneven distribution such as clumping, bare spots, or accumulation of solids along the slide edge.

6. If less than five slides remain, repeat Steps 1 through 5 to obtain five satisfactory slides.

7. Gram stain using the Hücker modification.

8. Count five slides at 1000X using oil immersion and normal illumination.

9. Use a microscope eyepiece graticule with a line ruled on it.

10. (a) Locate the scribe mark on the slide edge.
    (b) Line up the eyepiece line with the scribe mark on the slide.
    (c) Count any intersection with the eyepiece line of Gram-positive branched filaments of greater than 1 μm in length.
    (d) Move across the slide to the opposite edge, counting all intersections with the Gram-positive filaments greater than 1 μm in length.
    (e) Repeat steps (a) through (f) at the two other scribe marks on the slide.
    (f) Average the number obtained for the three counts and express the results as "number of intersections/g VSS."
    (g) Repeat procedure (a) through (f) for four more slides.

11. Average the results of (f) and (g).

12. The calculation is:

Nocardia Count (no. of intersections/g VSS) =

$$\frac{\text{Average no. of intersections}}{80 \ \mu\text{L}} \times \frac{1000 \ \mu\text{L}}{\text{mL}} \times \frac{1000 \ \text{mL}}{\text{liter}} \times \frac{1}{\text{MLVSS, g/L}}$$

>1 μm in a diluted Gram-stained sample of mixed liquor. The number of these filaments is assessed by counting the number of intersections they make with three equally spaced lines drawn across a stained microscope slide. In this sense the *Nocardia* counting technique is similar in concept to the simplified extended filament counting method used by the SJ/SCWPCP. *Nocardia* counts by this method are expressed as "Number of intersections per gram of VSS."

CHAPTER 2

Microscopic Examination of Activated Sludge with Special Reference to Floc Characteristics and Filamentous Organism Characterization

INTRODUCTION

Microscopic examination of activated sludge is useful for determining the physical nature of the activated sludge floc and the abundance and types of filamentous organisms present. This type of examination generally yields information related to the behavior of the activated sludge in solids separation and solids handling processes. This arises because it is the physical properties of the activated sludge revealed during microscopic examination that determine the settling and compaction characteristics of the sludge. Only a limited amount of information can be obtained on the biological or biochemical activity of activated sludge through such an examination. Judgments of whether an activated sludge is "healthy" or "unhealthy," "active" or "inactive," "young" or "old" based on a microscopic examination generally are not possible. In the rational diagnosis and rectification of activated sludge solids separation problems, microscopic examination and filamentous organism characterization is both necessary and invaluable.

This chapter will present an outline of the microscopic examination procedure, the types of information that can be obtained from each phase of this examination, the details of techniques used, and the identification or characterization of the types of filamentous organisms found. A section on interpretation of results, illustrated with specific case histories, is included.

SAMPLING

Activated sludge mixed liquor samples should be taken at points of good mixing either from the effluent end of the aeration basin or from the mixed liquor channel between the aeration basin and the secondary clarifier. Mixed liquor samples should be taken from below the surface to exclude any foam or other floating material. If the activated sludge is foaming, a separate foam sample should be taken from one of these points: the surface of the effluent end of the aeration basin, the surface of the mixed liquor channel, or the surface of the secondary clarifier. Subsurface liquid should be excluded from foam samples. Although foams can be thick and viscous and difficult to transfer to sample bottles, they should not be diluted for this purpose because one of the observations that may be important is the relative abundance of filamentous organisms in the foam compared to the mixed liquor.

Some activated sludge process modifications such as step-feed and contact stabilization have more than one aeration basin. In such systems the activated sludge in all basins usually has similar relative floc and filamentous organism characteristics. Thus it usually is not necessary to sample both basins—the effluent end of the aeration basin will suffice. One exception to this is in systems where the conditions in the various basins are different. For example, in activated sludge systems with an anaerobic/aerobic sequence of basins, some of the floc characteristics (specifically the presence of Neisser positive staining cells) may be different in the sludges in the two basins. Where an activated sludge plant consists of separate parallel systems there may be differences in floc and filamentous organism characteristics between the systems, especially if the RAS streams do not mix completely. In this situation a sample from each system is required. Similarly for two-stage activated sludge systems (e.g., "carbonaceous" first-stage activated sludge followed by "nitrification" second-stage activated sludge) a sample from the aeration basin of each stage is necessary for complete characterization of floc structure and filamentous organism populations. Some activated sludge systems treat a waste that has been pretreated in another type of biological treatment system, e.g., lagoon or trickling filter. These units may seed the activated sludge with filamentous organisms. To determine the extent of this, a sample of the wastewater entering the activated sludge system should be taken. Any side streams entering the aeration basin (e.g., anaerobic digester supernatant, aerobic digester decant, centrate) should be sampled to determine whether they are a source of filaments, or other materials in the flocs.

Sampling and examination frequency will be dictated by the circumstances and by where the samples are being examined. Daily examination can be justified for onsite examination during critical periods (e.g., when bulking sludge is occurring or is anticipated; during the use of RAS chlorination for bulking control; and during periods of experimental operation). For routine onsite examination, a frequency of about once every MCRT is generally adequate for routine activated sludge characterization. For offsite examination (e.g., by a laboratory skilled in the techniques), frequencies from weekly through monthly or seasonally have been employed, depending on the severity of problems being encountered or the desire to establish an operating history (and on the budget available for this activity).

SAMPLE TRANSPORT AND STORAGE

Samples should be examined as soon as possible. When examination is onsite, several hours delay is inconsequential; more lengthy storage should be at 4°C in a refrigerator. When samples must be transported offsite for analysis, they should be sent in sample containers in which there is an air space at least equal in volume to the sample volume to avoid septicity. A good and cheap container is a 5 mL plastic disposable transfer pipette. The bulb portion of the pipette should be filled about half full and then the tip should be heat sealed. The sample should be neither chemically preserved nor frozen since these procedures can alter the characteristics of both the flocs and filamentous organisms. The longer the time that elapses between sampling and examination, the more difficult and uncertain the sample examina-

tion and data interpretation become. Samples from plants with low F/M (high MCRT) maintain their characteristics longer than samples from plants with high F/M (low MCRT). For high MCRT samples, a satisfactory examination can be obtained if the sample is examined within 7 to 10 days of sampling; for sludges from low MCRT plants it is wise to examine the activated sludge within 3 to 4 days of its sampling. With the availability of express mail and overnight delivery services, these transit times should pose no problems for offsite analysis.

In our experience the most common mistakes that people make when sending samples by mail for analysis are:

- Failing to properly seal the container. Air travel involves pressure changes that encourage sample leakage. It is very difficult to analyze a sample that has been scraped half dry from the inside of an envelope (and the delivery service does not appreciate it either!!).
- Filling the sample bottle completely full. Besides encouraging leakage due to pressure changes, this often results in septicity that can influence some filamentous organism and floc characteristics.
- Sending too much sample. When we receive a 5-gallon container completely full (and dripping) with septic activated sludge, we wonder whether the sender wants an activated sludge microscopic analysis or whether they think they have discovered a new (though not very cost-effective) method of sludge disposal.
- Sending samples in ice chests with ice or "magic-cold" type coolers. Again, this is an unnecessary expense.

Filamentous organism staining reactions can be quite sensitive to prolonged sample storage. If long periods between sampling and examination are anticipated, two air-dried smears on microscope slides should be prepared at the time of sampling (using the techniques outlined later) and sent together with the sample. In this way the original characteristics of the activated sludge will be preserved for Gram and Neisser staining. The slides should be marked with sample identification, date, and G or N (Gram or Neisser) on the same side of the slide that the smear was made. The same procedure should be followed if samples cannot be analyzed immediately upon receipt.

EQUIPMENT AND SUPPLIES REQUIRED

The central tool is the microscope. A research grade phase contrast microscope with 100X and 900–1000X phase contrast objectives is required. Phase contrast objectives yielding 200X and 400X magnification are useful but not essential. A mechanical stage is essential for controlled scanning. A binocular head (or trinocular head, if photography is desired) and built-in light source are highly recommended. An ocular scale should be inserted into one of the eyepieces and calibrated at each magnification with a stage micrometer. Microscopes with the features described above currently cost about $3,500. An example of an excellent microscope for activated sludge examination is shown in Figure 11.

A photographic record of sample appearance at 100X is recommended for reference purposes. A 35 mm camera mounted on the trinocular head is most convenient. The camera should have a built-in light meter and a replaceable, fine ground-glass focusing screen. If you want to go "first-cabin," many microscope manufacturers make automatic cameras specifically for microphotography. To do things more economically, use a Polaroid instant film camera. A convenient and highly efficient system is to take all photographs using color slide film (e.g., Kodak KR-64 for daylight balanced light sources and Kodak KPA for tungsten balanced light sources). Color or black and white slides or prints can be produced from these using a slide printer with Polaroid instant film.

Since many of the observations required are close to the limit of resolution of the light microscope, and because many of the features observed are difficult to detect, the microscope must be in tip-top condition. Regular adjustment of the phase rings is required; the microscope objectives must be kept clean; and a dust-free environment is necessary. Professional servicing of the microscope by the manufacturer or their recommended agent is recommended at least once per year.

Two staining procedures are used routinely as part of the floc and filamentous organism characterization procedure—these are the Gram stain and the Neisser stain. Additional procedures with more specialized uses are a test for sulfide oxidation to intracellular sulfur granules (the "S test"); an India ink reverse stain to detect the presence of large amounts of exocellular polymeric materials ("slime") in the floc and on the filament surfaces; a staining procedure for the detection of intracellular storage products such as poly-β-hydroxybutyrate (PHB); and a sheath stain done on the sample wet mounts using a 0.1% aqueous Crystal Violet solution. The reagents and techniques required for each of these tests are presented in Tables 7–13. These tables also include a description of the anthrone test for determining the carbohydrate content of activated sludge; this test can be used for additional confirmation of the presence of large amounts of exocellular polymer, observed microscopically by India ink reverse staining.

MICROSCOPIC EXAMINATION PROCEDURE

Upon sample receipt, or at least within several hours prior to sample examination, spread one drop of well-mixed sample evenly over approximately 50% of the area of each of two 25 mm × 75 mm microscope slides. Allow these slides to air dry at room temperature (**do not heat fix**). The slides can be stored indefinitely and stained later.

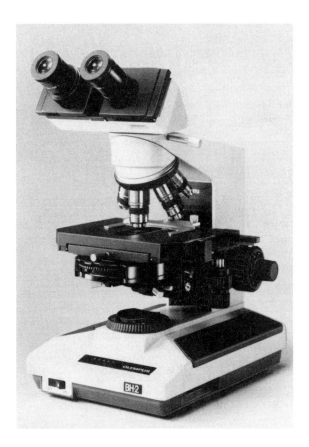

Figure 11. A modern binocular phase contrast microscope.

Table 7. Gram Stain, Modified Hücker Method

Reagents   Prepare fresh every 3–6 months.

Solution 1   Prepare the following separately, then combine

A		**B**	
Crystal Violet	2 g	Ammonium oxalate	0.8 g
Ethanol, 95%	20 mL	Distilled water	80 mL

Solution 2

Iodine	1 g
Potassium iodide	2 g
Distilled water	300 mL

Decolorizing Solution

Ethanol, 95%

Solution 3

Safranin O (2.5% w/v in 95% ethanol)	10 mL
Distilled water	100 mL

Procedure

1. Prepare thin smears on microscope slides and thoroughly air dry (do not heat fix).

2. Stain 1 min with Solution 1; rinse 1 sec with water.

3. Stain 1 min with Solution 2; rinse well with water.

4. Hold slide at an angle and decolorize with 95% ethanol added drop by drop to the smear for exactly 25 sec. **Do not over decolorize**. Blot dry.

5. Stain with Solution 3 for 1 min; rinse well with water and blot dry.

6. Examine under oil immersion at 1000X magnification with direct illumination (not phase contrast): blue-violet is positive; red is negative (Plate 1).

Mark the slides with a sample identification code and a G or N (Gram or Neisser) on the side on which the smear has been made (frosted-end slides are recommended). Perform the Gram and Neisser staining procedures outlined in Tables 7 and 8.

Withdraw one drop (approximately 0.05 mL) of well-mixed sample with a loop or a clean, disposable Pasteur pipette and place on a 25 mm × 75 mm microscope slide. Place a 22 mm No. 1 glass coverslip on the drop and press down very gently on the coverslip with a blunt object. Remove the liquid that is expelled from the sides of the coverslip with a tissue. This procedure produces the thin preparation which is necessary because of the limited depth of focus of the microscope optical system. This is necessary, even though it hastens the drying out of the slide, because at high power observation, long filaments can pass in and out of focus along their length.

When pressing down, physical resistance or cracking of the coverslip is an indication that the sample contains some fairly large solid (nonbiological) particles. Some examples encountered have included activated carbon, resin beads, and pieces of ceramic.

Examine the wet mount under phase contrast illumination at 100X magnification for the following characteristics:

1. The general size and shape of flocs; measure approximately 10–20 flocs and place them in the following categories:

Table 8. Neisser Stain (Eikelboom and van Buijsen, 1981)

Reagents Prepare fresh every 3–6 months.

Solution 1

Separately prepare and store the following

A		**B**	
Methylene Blue	0.1 g	Crystal Violet (10% w/v in 95% ethanol)	3.3 mL
Ethanol, 95%	5 mL	Ethanol, 95%	6.7 mL
Acetic acid, glacial	5 mL	Distilled water	100 mL
Distilled water	100 mL		

Mix 2 parts by volume of A with 1 part by volume of B.

Solution 2

Bismark Brown , $C_{18}H_{18}N_8$ (1% w/v aqueous)	33.3 mL
Distilled water	66.7 mL

Procedure

1. Prepare thin smears on microscope slides and thoroughly air dry (do not heat fix).

2. Stain 30 sec with Solution 1; rinse 1 sec with water.

3. Stain 1 min with Solution 2; rinse well with water; blot dry.

4. Examine under oil immersion at 1000X magnification with direct illumination (not phase contrast): blue-violet is positive (either entire cell or intracellular granules); yellow-brown is negative (Plate 2).

---

  a. Floc size range (maximum dimension or diameter if approximately spherical)
  small ≤150 μm
  medium 150–500 μm
  large ≥500 μm

  b. Floc shape, whether round or irregular, whether compact or diffuse; also note whether texture is firm, or weak (Figure 12).

2. The presence and types of protozoa and other macroorganisms (e.g., rotifers, nematodes) (Figures 36–40).

3. The presence of nonbiological organic and inorganic particles (Figure 34).

4. The presence of fingered or amorphous zoogloeal colonies (Figure 12c and d).

5. The presence of free-floating (dispersed) cells in the bulk solution (Figure 12b). The supernatant of samples containing significant amounts of dispersed cells will appear turbid.

6. The presence and effect of filamentous organisms on floc structure:
  a. None

  b. Bridging—the filaments extend from the floc surface into the bulk solution, and bridge between the flocs (Figures 5a and b).

  c. Open floc structure—the floc population attaches to and grows around the filamentous organisms leading to large, irregularly-shaped flocs with substantial internal voids. (Figures 5c and d).

7. The abundance of filamentous organisms. This can be measured by several methods. For this type of analysis a subjective scoring system is used. Filamentous organisms are observed first at 100X and then at 1000X and on the basis of these observations are subjectively rated for overall abundance on a scale from 0 (none) to 6 (excessive) (Table 14). An abundance rating is determined for the sample as a whole (all filament types together) and for each filamentous organism observed. Individual filamentous organisms are considered dominant (and likely most responsible for solids separation problems) if they are scored "very com-

Table 9. Sulfur Oxidation Test (S Test)

Test A   (modified from Eikelboom and van Buijsen, 1981)

Reagent

Sodium sulfide solution ($Na_2S.9H_2O$) 1.0 g/L; prepare fresh weekly.

Procedure

1. On a microscope slide mix 1 drop of activated sludge sample and 1 drop of sodium sulfide solution.

2. Allow to stand open to the air for 10–20 min.

3. Place a coverslip on the preparation and gently press to exclude excess solution; remove expelled solution with a tissue.

4. Observe at 1000X using phase contrast. A positive S test is the observation of highly refractive, yellow-colored intracellular granules (sulfur granules) (Figure 20 and Plate 3e).

This test can give variable results. This is due to methodological problems involving the relative concentrations of sulfide and oxygen present (sulfide oxidation is an aerobic process). An alternative sulfur oxidation test (Test B), developed by Nielsen (1985), may be used.

Test B   (modified from Nielsen, 1985)

Reagent

Sodium thiosulfate ($Na_2S_2O_3.5H_2O$) 1 g/100 mL; prepare fresh weekly.

Procedure

1. Allow activated sludge sample to settle, and transfer 20 mL of clear supernatant to a 100 mL Erlenmeyer flask.

2. Add 1–2 mL of mixed activated sludge sample to the flask.

3. Add 1 mL of sodium thiosulfate solution to the flask (final sodium thiosulfate concentration is 2mM).

4. Shake the flask overnight at room temperature.

5. Observe at 1000X phase contrast, as in Test A.

mon" or greater. Organisms are considered secondary, i.e., they are present but not in sufficient abundance to account for solids separation problems, if they are scored "common" or less. This method is both rapid and suitable for establishing whether a filamentous organism is dominant or secondary and for determining their response to remedial actions. Abundance categories generally are reproducible to within ± one abundance category between observers. With practice it is possible to achieve a consistent relationship between the subjective scoring of filament abundance and measures of sludge settleability such as SVI. For example, Figure 13 shows data taken from an industrial wastewater activated sludge system. SVI was measured at the wastewater treatment plant and the subjective scoring of filament abundance was done by an independent laboratory. Pictorial examples of filament abundance categories are shown in Figure 14.

OBSERVATION OF FILAMENTOUS ORGANISM CHARACTERISTICS

The next step is to determine which types of filamentous organisms are present and their individual abundances. Change the microscope to the 1000X phase contrast setting and carefully char-

Table 10. India Ink Reverse Stain

Reagent

India ink (aqueous suspension of carbon black particles).

Procedure

1. Mix one drop of India ink and one drop of activated sludge sample on a microscope slide.[a] Depending on the ink used, the sample volume may need to be reduced.

2. Place on the coverslip and observe at 100X using phase contrast.

3. In "normal" activated sludge, the India ink particles penetrate the flocs almost completely, at most leaving a clear center (Figure 4b).

4. In activated sludge containing large amounts of exocellular polymeric material, there will be large, clear areas which contain a low density of cells (Figures 4d and e).

[a]Alternatively, apply one drop of India ink to the edge of a coverslip under which a thin preparation of the sample has been prepared. This technique allows the observer to see the India ink particles spreading across the slide. When using this method be careful not to contaminate the immersion oil on the coverslip with India ink.

Table 11. Poly-β-Hydroxybutyrate (PHB) Stain[a]

Reagents

Solution 1   Sudan Black B (IV), 0.3% w/v in 60% ethanol.

Solution 2   Safranin O, 0.5% w/v aqueous solution.

Procedure

1. Prepare thin smears on a microscope slide and thoroughly air dry.

2. Stain 10 min with Solution l; add more stain if the slide starts to dry out.

3. Rinse 1 sec with water.

4. Stain 10 sec with Solution 2; rinse well with water; blot dry.

5. Examine under oil immersion at 1000X magnification with direct illumination (not phase contrast). PHB granules will appear as intracellular, blue-black granules, while the cytoplasm will be pink or clear.

[a]*Source: Manual of Methods for General Microbiology*, R.G.E. Murray Ed., American Society for Microbiology, Washington, D.C., 1981.

Table 12. Crystal Violet Sheath Stain

Reagent   Crystal Violet, 0.1% w/v aqueous solution.

Procedure

1. Mix 1 drop of activated sludge sample and 1 drop of Crystal Violet solution on a microscope slide, cover and examine at 1000X magnification using direct illumination. The cells stain deep violet, while the sheaths are clear to pink in color.

Table 13. Anthrone Test for Carbohydrate[a]

Carbohydrates when boiled with concentrated sulfuric acid form furfurals (from both pentoses and hexoses) which react with aromatic amines to form colored products. These products can be measured spectrophotometrically.

Reagents

Sulfuric acid	75% v/v reagent grade. Add 750 mL concentrated sulfuric acid to 250 mL distilled water (be sure to add the acid to the water)! Cool to room temperature before using.
Anthrone reagent	Add 200 mg anthrone (9 [10H]-anthracenone; CAS. No. 90–44–8; $C_6H_4COC_6H_4CH_2$) to 5 mL absolute ethanol; make up to 100 mL with 75% sulfuric acid. Store at 5°C and replace monthly or sooner if a brown color develops.
Glucose standard	Add 100 mg glucose to 100 mL distilled water and add 150 mg benzoic acid (preservative). Store at 5°C. Dilute 1:10 with distilled water for daily use (1 mL = 100 μg glucose).
Sodium chloride	Dissolve 8.5 NaCl in 1000 mL distilled water (0.85% solution).
Equipment	Thick-walled Pyrex boiling tubes (15 cm × 2.5 cm) Boiling water bath Ice water bath Spectrophotometer (625 nm) and cuvettes

Procedure

1. Centrifuge an activated sludge mixed liquor sample of known MLSS content; wash with 0.85% NaCl solution; recentrifuge; resuspend to original volume with distilled water. **Note:** this step may be omitted if the results from washed and unwashed samples are the same.

2. Pipette a range of sample volumes from 0.1 to 1.0 mL into a series of boiling tubes; adjust all final volumes to 1.0 mL with distilled water.

3. Using the 100 μg/mL glucose standard and distilled water, prepare 3–4 glucose standards in the range of 10–100 μg glucose each in water to a final volume of 1.0 mL (0.1 to 1.0 mL of the 1:10 glucose standard per tube). Include a 1.0 mL distilled water blank.

4. Chill all tubes in the ice water bath.

5. Add 5.0 mL chilled anthrone reagent to each tube; mix thoroughly and keep these in the ice water bath.

6. Transfer all tubes to the boiling water bath for exactly 10 min.

7. Return the tubes to the ice water bath.

8. Measure the absorbance of all tubes at 625 nm using the distilled water tube as a blank.

9. Prepare a standard curve for the glucose standard by plotting log % transmittance versus glucose concentration. This should be a straight line. Read the sample glucose content from this standard curve. Use only sample values that are between ≤90% transmittance and the % transmittance produced by the highest glucose standard that lies on the straight line portion of the standard curve. Express results as μg/mL glucose. Convert this to mg "total carbohydrate" per gram dry weight activated sludge, or report as % of dry weight as glucose.

NOTE: after some experience with this test, the proper dilutions of the activated sludge can be determined and the number of tests reduced.

[a]*Source:* Modified from *Manual of Methods for General Bacteriology,* R.G.E. Murray, Ed., American Society for Microbiology, Washington, D.C., 1981.

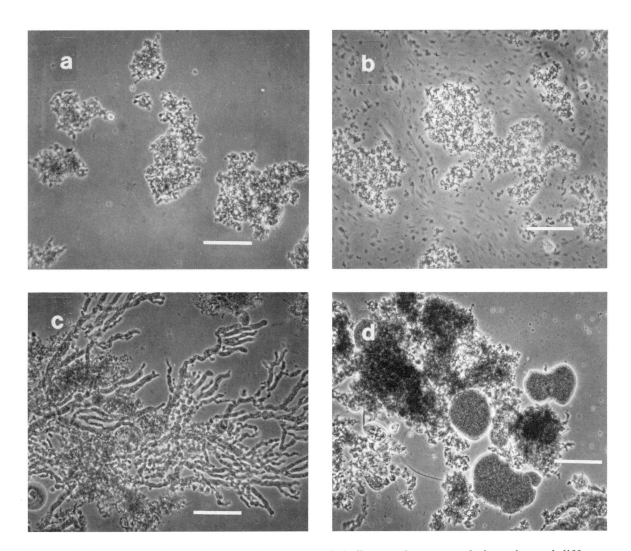

Figure 12. Floc "texture" in activated sludge: *a.* rounded, firm, and compact; *b.* irregular and diffuse with substantial free cells. Appearance of zoogloeal organisms in activated sludge: *c.* fingered; *d.* amorphous. (all 100X phase contrast; bar = 100 μm).

acterize each filamentous organism present by looking at several filaments of each type and expressing the results as an "average." Use filamentous organism worksheets such as the ones presented in Tables 15 and 16 to record and summarize individual observations.

This task can be simplified by accepting only a limited number of descriptions, and learning to recognize special features that provide clues to the filamentous organism type. The characteristics are:

1. Branching: present or absent; if present, whether true or false.

   True branching refers to cell branching where there is contiguous cytoplasm between branched trichomes (Figures 15a, b, and c). In activated sludge the only trichome-forming organisms that commonly have true branches are fungi and *Nocardia* spp. *Nostocoida limicola* also is very rarely observed with a true branch. False branching occurs when there is no contiguous cytoplasm between trichomes; two trichomes have merely stuck together and grown outward (Figure 15d). In activated sludge, false branching is only observed for *Sphaerotilus natans*. Type 1701 has been observed to exhibit false branching, but only in pure culture.

2. Motility: none, or, if present, describe.

   Only a few filamentous organisms in

Table 14. Subjective Scoring of Filament Abundance[a]

Numerical Value	Abundance	Explanation
0	none	
1	few	Filaments present, but only observed in an occasional floc
2	some	Filaments commonly observed, but not present in all flocs
3	common	Filaments observed in all flocs, but at low density (e.g., 1–5 filaments per floc)
4	very common	Filaments observed in all flocs at medium density (e.g., 5–20 per floc)
5	abundant	Filaments observed in all flocs at high density (e.g., >20 per floc)
6	excessive	Filaments present in all flocs—appears more filaments than floc and/or filaments growing in high abundance in bulk solution

[a] Note: this scale from 0 to 6 represents a 100–1000 fold range of total extended filament length.

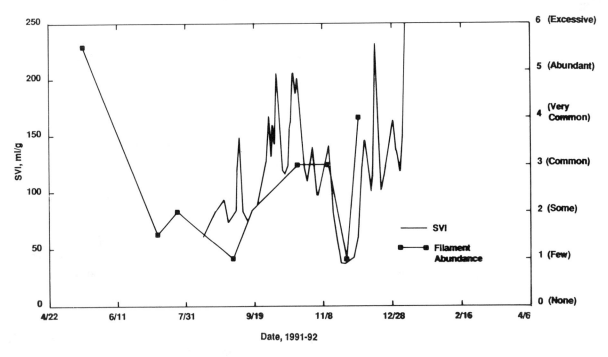

Figure 13. Relationship between subjective scoring of filament abundance and SVI as measured by independent observers.

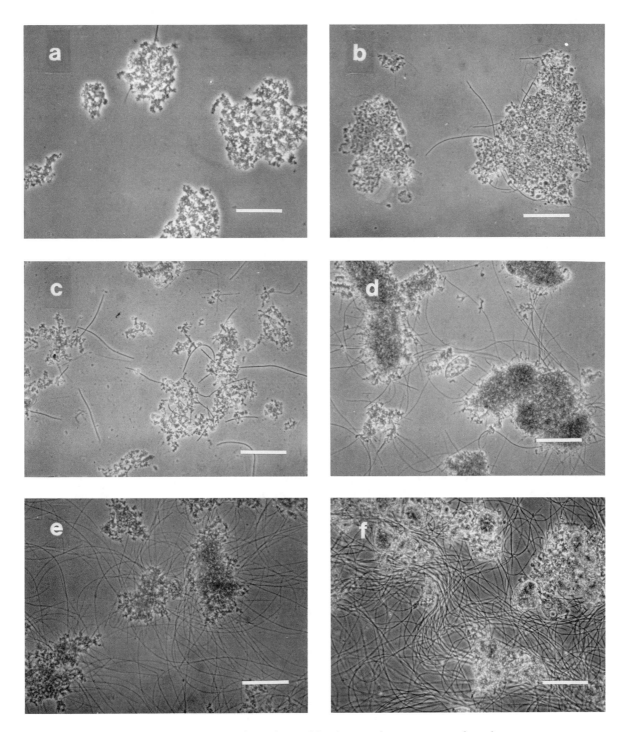

Figure 14. Filament abundance categories using subjective scoring system: *a.* few; *b.* some; *c.* common; *d.* very common; *e.* abundant; and *f.* excessive (all 100X phase contrast; bar = 100 μm).

Table 15.   Floc Characterization/Filamentous Organism Identification Worksheet—Page One

Sample Number_____   Sample Location_____

Sample Date___/___/___   Observation Date___/___/___

Filament Abundance: ☐   ☐   ☐   ☐   ☐   ☐   ☐

                      0   1   2   3   4   5   6

          None   Few   Some   Common   Very   Abundant   Excessive

                                                Common

Filament Effect on Floc Structure: ☐ Little or None   ☐ Bridging   ☐ Open Floc Structure

Morphology of Floc: ☐ Firm   ☐ Weak   ☐ Round   ☐ Irregular   ☐ Compact   ☐ Diffuse

Size (μm): | <150 | 150–500 | >500 |
(% in range) | | | |

Features: Free cells in suspension   _____   Neisser positive-cell clumps   _____
            Zoogloeas   _____   India Ink Test   _____
            Spirochaetes   _____   Chlorine damage to filaments_____
            Inorganic/Organic Particles_____

FILAMENTOUS MICROORGANISM SUMMARY:

	Rank	Abundance		Rank	Abundance
Nocardia spp.			*M. parvicella*		
Type 1701			Type 0581		
S. natans			Type 0092		
Type 021N			Type 0803		
Thiothrix spp.			Type 1851		
Type 0041			Type 0691		
H. hydrossis			Type 0675		
N. limicola			Other		
Fungus			Other		

Remarks_____

_____

_____

_____

_____

_____

Table 16. Floc Characterization/Filamentous Organism Identification Worksheet – Page Two

Sample Number_____   Sample Location_____

Sample Date____/___/___   Observation Date____/___/___

COMMENTS

OBSERVATION OF
    Protozoa
    Metazoa

WET MOUNT OBSERVATION; 1000X, PHASE CONTRAST FOR FILAMENT
CHARACTERISTICS; 1000X BRIGHT FIELD FOR STAINS

FILAMENT NO.	A	B	C	D	E
Branching					
Motility					
Filament Shape					
Filament Location					
Attached Growth					
Sheath					
Crosswalls					
Filament Diameter, μm					
Filament Length, μm					
Cell Shape					
Cell Size, μm					
Sulfur Deposits					
Other Granules					
Gram Stain					
Neisser Stain					
Commonness					
Rank					
Identification					

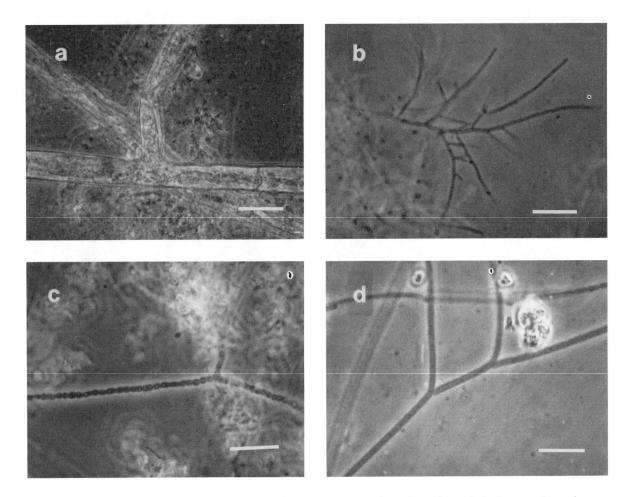

Figure 15. Trichome branching observed for filamentous organisms in activated sludge: *a.*, *b.* and *c.* true branching (fungus, *Nocardia* spp., and *Nostocoida limicola* II, respectively); *d.* false-branching (*Sphaerotilus natans*) (all 1000X phase contrast; bar = 10 μm).

activated sludge are motile. *Beggiatoa* spp., *Flexibacter* spp. and some blue-green bacteria (*Cyanophyceae*) are motile by gliding; *Thiothrix* spp. and type 021N may display limited "twitching" or swaying motions.

3. Filament shape: straight, smoothly-curved, bent, irregularly-shaped "chain of cells," coiled, or mycelial (Figure 16).
4. Location: extending from floc surface, found mostly within the floc, or free-floating in the liquid between flocs (Figures 17d, e, and f).
5. Attached growth of epiphytic bacteria: present or absent; if present, whether the growth is substantial or incidental (Figures 17a, b, and c).
6. Sheath: present or absent.

The presence of a sheath is one of the most difficult characteristics to establish.

Since a true sheath is a clear structure exterior to the cell wall, it is difficult to see. Sheaths can be seen best in unstained preparations when they are empty, i.e., they do not contain cells, or when some of the cells are missing. In the latter case the outline of the sheath can be seen continuing along either side of the empty space (Figure 18). Several features of filamentous organisms can be confused with sheaths. A yellowish "halo" observed around filaments viewed under phase contrast illumination is not a sheath. It is an artifact of the phase contrast illumination. Short, empty spaces in a trichome or at the trichome apex should not be used to indicate the presence of a sheath. The cell wall of some filamentous organisms may remain after cell lysis; however, this can be distinguished from a

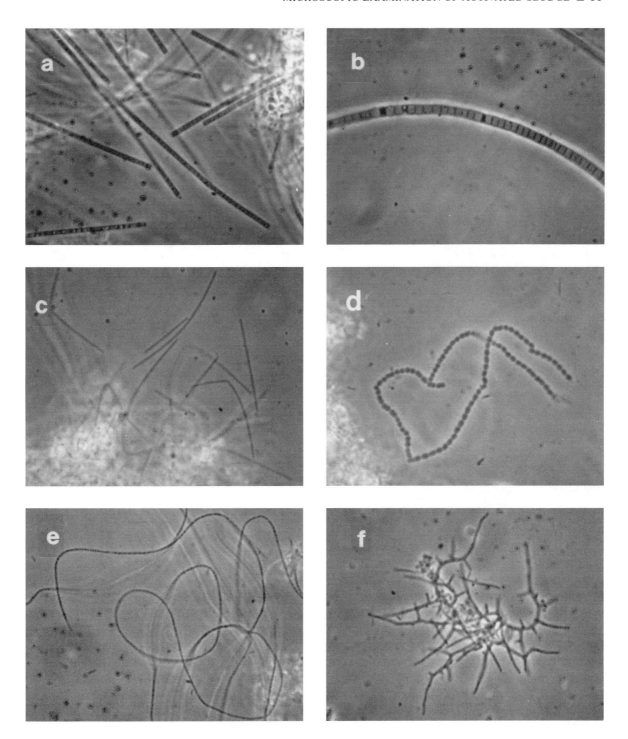

Figure 16. Examples of filament shapes: *a.* straight; *b.* smoothly curved; *c.* bent; *d.* irregularly-shaped "chain of cells"; *e.* coiled; and *f.* mycelial (all 1000X phase contrast).

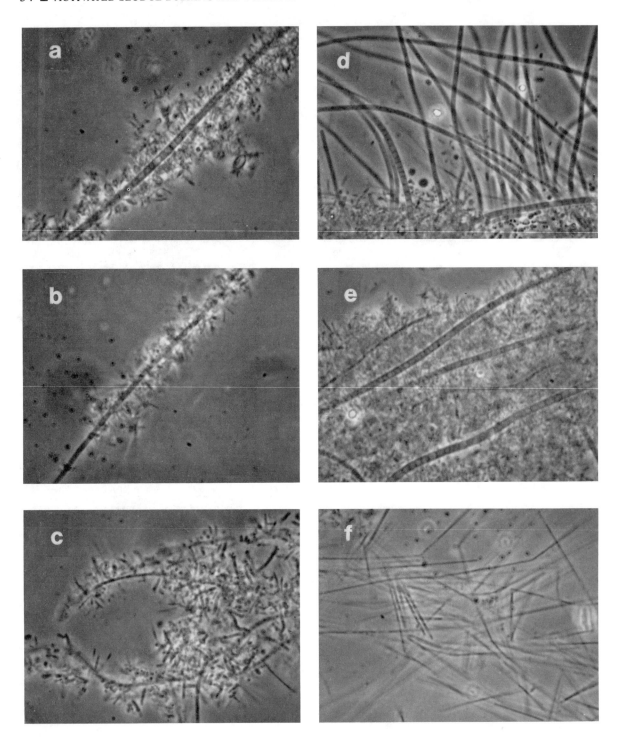

Figure 17. Attached growth of epiphytic bacteria on filamentous organisms: *a*. type 0041; *b*. type 0675; *c*. type 1701. Location of filaments in activated sludge: *d*. extending from floc surface; *e*. mostly within the floc; *f*. free (all 1000X phase contrast).

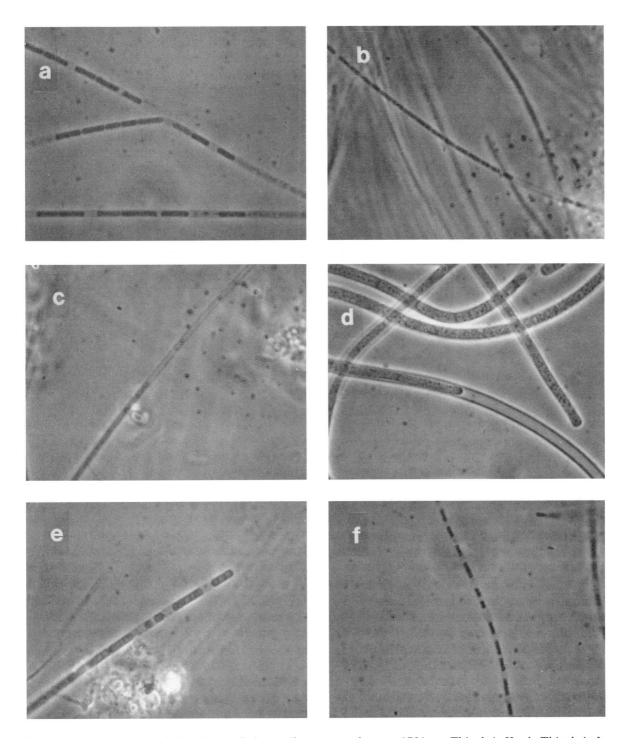

Figure 18. Appearance of sheaths: *a. Sphaerotilus natans*; *b.* type 1701; *c. Thiothrix* II; *d. Thiothrix* I; *e.* type 0041; and *f.* type 1701, stained with Crystal Violet (all 1000X phase contrast).

sheath because some evidence of pre-existing cross-walls usually remains. This commonly is observed for type 021N.

The presence of substantial attached bacterial growth (epiphytic growth) generally indicates the presence of a sheath. Staining wet mounts with Crystal Violet (Table 12) may aid in sheath detection (Figure 18f). Sheaths also can be visualized by mixing equal volumes of activated sludge and a 1/1000 dilution of household bleach (sodium hypochlorite solution) and letting the mixture stand for several hours or overnight. This treatment lyses the cells inside the sheath, leaving empty sheaths that can be seen more easily under phase contrast illumination at 1000X.

7. Cross-walls (cell septa): present or absent.

   This feature can be variable for some filamentous organisms and detection depends on the quality and adjustment of the microscope. It is important to determine whether a true trichome is present (Figure 16b) or whether the filament is made up of a chain of cells (Figure 16d).

8. Filament diameter: both the average diameter and its range (in μm) should be measured. It is important to note whether the average diameter is greater or less than 1 μm.

9. Filament length: report the range in μm.

10. Cell shape: square, rectangular, oval, barrel, discoid (stacked "hockey pucks"), round-ended rods ("sausage-shaped") or irregular (Figure 19).

    It is important to note whether there are indentations at cell septa (Figures 19c, d, e, and f) or whether trichome walls are straight at the cell junctions (Figures 19a and b).

11. Size: report the average length and width of the cells in μm.

12. Sulfur deposits: present or absent in situ and present or absent after performing the S test (Table 9), (Figures 20a and b).

    Under phase contrast observation sulfur granules appear as bright, refractive, yellow-colored cell inclusions (Plate 3e), either in the shape of spheres observed for *Thiothrix* spp., *Beggiatoa* spp., and type 021N (Figures 20c, d, and e), or in the shape of squares or rectangles for type 0914 (Figure 20f). Type 0914 does not respond to the S test.

13. Other granules: present or absent. Commonly observed granules are polyphosphate (Neisser positive granules; Plates 3a and b), and PHB (confirmed by PHB staining) (Table 11 and Plate 3f).

14. Staining reactions: each filamentous organism present is separately evaluated for Gram staining and Neisser staining reactions by observing the stained, air-dried smears at 1000X using direct illumination (not phase contrast). The general location and length of filamentous organisms in the wet mount, and the presence or absence of attached growth should be noted carefully so that the same filament types can be examined in the stained smears. Care is required in this observation because some filamentous organisms change size upon drying and staining (e.g., type 0092 appears much wider when Neisser stained than in wet mounts).

    The Gram stain (Table 7) requires much practice. Reagents should be reasonably fresh (3–6 months) and, if possible, should be tested on fresh cultures with known Gram staining reactions. The decolorization step should be controlled precisely to avoid over-decolorization. Add the decolorizer solution dropwise — not in a continuous stream — and add it for exactly 25 sec. Large dense flocs do not decolorize fully so the Gram reactions inside them should be ignored.

    Score the Gram reaction as strongly positive, weakly positive, variable, or negative. Most filamentous organisms observed in activated sludge are Gram negative (Plates 1a and b). *Nostocoida limicola*, types 0041 and 0675 most often are Gram positive but can be Gram variable or Gram negative (Plate 1c). Type 1851 stains weakly Gram positive, and generally appears as a chain of Gram positive "beads" (Plate 1d). *Thiothrix* I, *Beggiatoa* spp., type 021N and type 0914 generally stain Gram negative, but may stain Gram positive when they contain substantial intracellular sulfur deposits. *Microthrix parvicella* and *Nocardia* spp. are generally strongly Gram positive (Plates 1e and f).

    Neisser staining (Table 8) is a straightforward technique. Score as negative, positive (entire trichome is stained), or negative with Neisser-positive granules (Plate 2).

    Type 0092 (light purple — Plate 2c) and *N. limicola* (dark purple — Plate 2d) stain

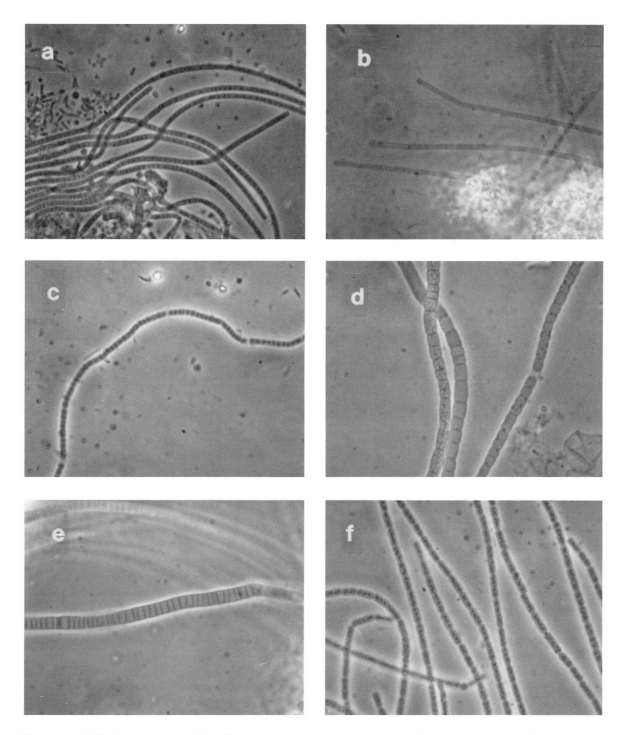

Figure 19. Cell shapes observed for filamentous organisms in activated sludge: *a*. square; *b*. rectangular; *c*. oval; *d*. barrel; *e*. discoid; and *f*. round-ended rods (all 1000X phase contrast).

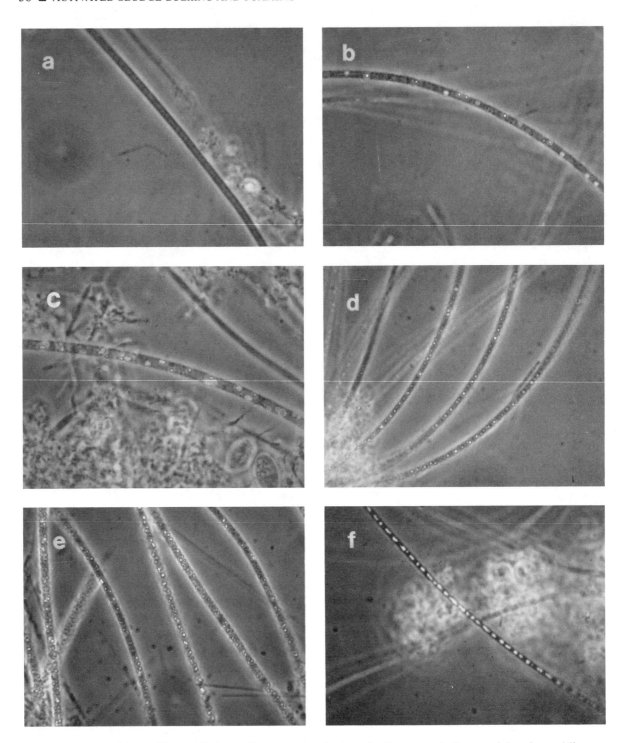

Figure 20. Deposition of intracellular sulfur granules during the S test: *a.* before and *b.* after adding sodium sulfide. Appearance of intracellular granules in filamentous organisms: *c: Thiothrix* I; *d. Thiothrix* II; *e.* type 021N; and *f.* type 0914 (all 1000X phase contrast).

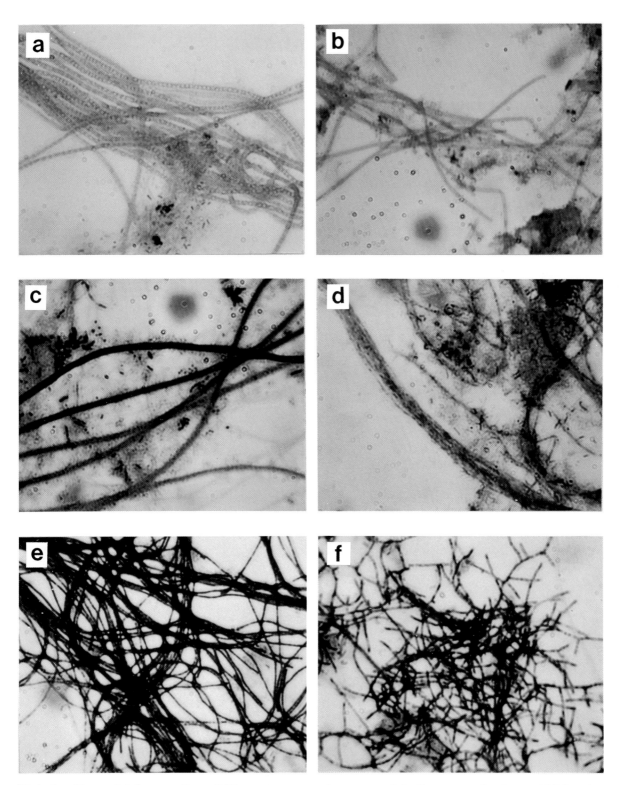

Plate 1. Gram-staining reaction of filamentous organisms: *a*. and *b*. Gram negative (types 021N and 0092, respectively); *c*. Gram variable (type 0041); *d*. weakly Gram positive (type 1851); and *e*. and *f*. Gram positive (*Microthrix parvicella* and *Nocardia* spp., respectively) (all 1000X direct illumination light).

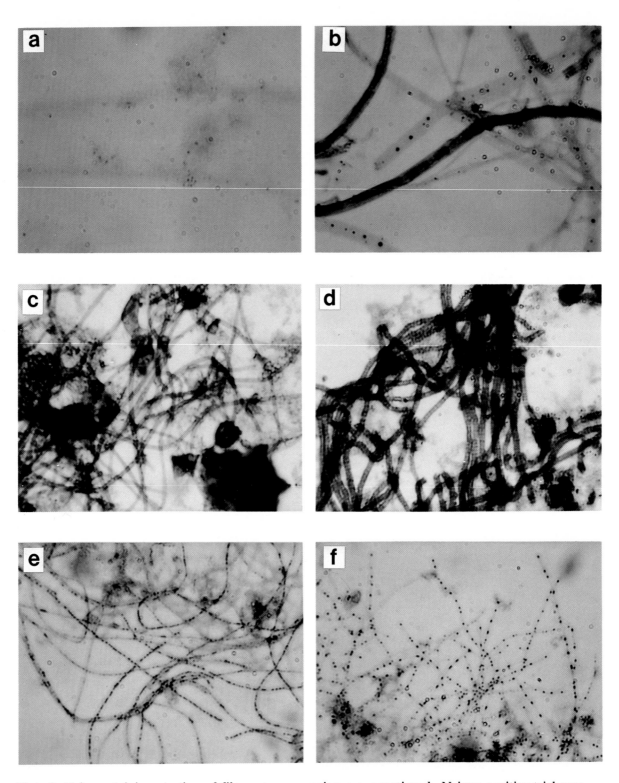

Plate 2. Neisser staining reaction of filamentous organisms: *a*. negative; *b*. Neisser positive trichome covering observed atypically for type 0041; *c*. and *d*. Neisser positive (type 0092 and *Nostocoida limicola* II, respectively); and *e*. and *f*. Neisser positive granules (*Microthrix parvicella* and *Nocardia* spp., respectively) (all 1000X direct illumination light).

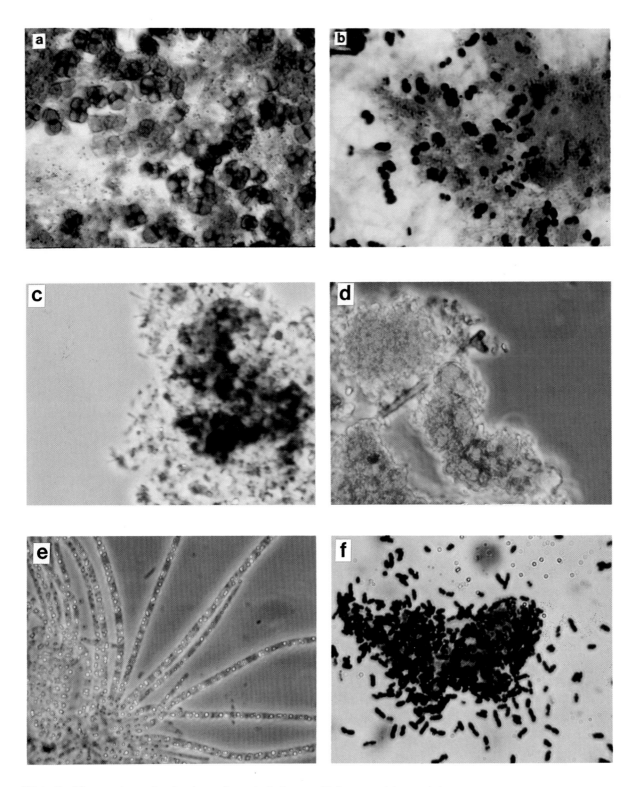

Plate 3. Observation of color in activated sludge: *a*. Neisser-positive staining tetrads; *b*. Neisser-positive staining bacteria in EBPR systems; *c*. precipitated sulfide in the floc; *d*. precipitated iron in the floc; *e*. intracellular sulfur granules in *Thiothrix* spp. *f*. PHB-positive staining granules in activated sludge bacteria. (*a.*, *b.* and *f.*, direct illumination light; *c.*, *d.* and *e.* phase contrast; all 1000X).

entirely Neisser positive. *M. parvicella* and *Nocardia* spp. usually stain Neisser negative but generally contain Neisser positive intracellular granules (Plates 2e and f). *Beggiatoa* spp., *Thiothrix* spp., and types 0041, 0675, 021N, 0914, and 1863 may infrequently contain Neisser positive granules. In addition, *H. hydrossis* and types 0675 and 0041 may have a Neisser positive trichome "covering" (Plate 2b) when present in industrial wastewater activated sludge systems.

15. Additional observations: two filamentous organisms, *Thiothrix* spp. and type 021N (uncommonly), may display rosettes and gonidia. A rosette develops when trichomes radiate outward from a common origin (Figures 26b, c, and e). Gonidia are oval- or rod-shaped cells present at the trichome apex that are distinctively different in appearance from the rest of the (vegetative) cells further down the filament (Figures 25e and 26f). Rosettes and gonidia indicate that the organisms are growing rapidly. They are also seen in nutrient deficient conditions, as well as when a septic wastewater is being treated.

FILAMENTOUS ORGANISM "IDENTIFICATION"

The observed and recorded filament characteristics entered on the Floc Characterization and Filament Identification Worksheet (Tables 15 and 16) are used to characterize the filamentous organisms to genus or to a numbered "type" using the dichotomous key shown in Figure 21. This procedure is simplified in two ways. First, only a limited number of characteristics are accepted in describing the organisms (Tables 16 and 17) and second, the key only lists the 22 filamentous bacteria most commonly observed in activated sludge. The infrequently observed filament types 1702, 1852, and 0211 are omitted. To further simplify the key, several filamentous organisms having readily identifiable specific characteristics are not included but are described separately. These include: fungi, *Cyanophyceae*, *Flexibacter* spp., and *Bacillus* spp.

This dichotomous key is a modification of the filamentous organism identification key given by Eikelboom and van Buijsen (1981), with changes to de-emphasize the need for the observation of cell septa (cross-walls), which can depend on the

quality and adjustment of the microscope used; and to include some filamentous organisms in the key twice where an important characteristic is variable, e.g., Gram stain reaction for *N. limicola* II and the observation of intracellular sulfur granules for types 0914 and 021N.

The use of this key is not without risk because some filamentous organism characteristics vary, and the key cannot always address all variables. The filament type arrived at using the key should be checked carefully against the typical organism characteristics listed in Table 17 and presented in the short descriptions and photographs of each organism that follow. If the characteristics given in Table 17, or shown in the photographs, do not correspond to the filament type arrived at using the key, the characteristics used in the key should be re-examined carefully. For example, type 0041 usually gives a weak Gram positive or a Gram variable reaction, and, as such, is keyed correctly from Figure 21. However, if strongly Gram positive, it would be keyed as *N. limicola* II. Reference to Table 17 shows *N. limicola* II to be coiled and to be free of attached growth, while type 0041 is straight or smoothly-curved and most often has substantial, attached growth. Here, the Gram-stained slide should be re-examined.

Occasionally, a filamentous organism is observed that cannot be placed in a type or genus designated in the key. The organism should be reported as "not identified"; do not try to "force fit" the organism into existing filament types.

Probably the hardest part of filamentous organism identification is lacking the confidence to make an identification. The best ways to overcome this are to:

1. identify the filamentous organisms in the sample, then send the same sample to someone skilled in the technique. Many groups of treatment plants in a particular area establish round-robin filament identification procedures to assist each other;
2. take a course in the techniques of microscopic activated sludge characterization and filamentous organism identification.

For someone located at a particular treatment plant, the task is not as awesome as it might first appear. It is doubtful that, in such a situation, you will see all of the filamentous organism types listed below. Usually, you will encounter up to about half a dozen. . . and soon you will learn to recognize them easily!!

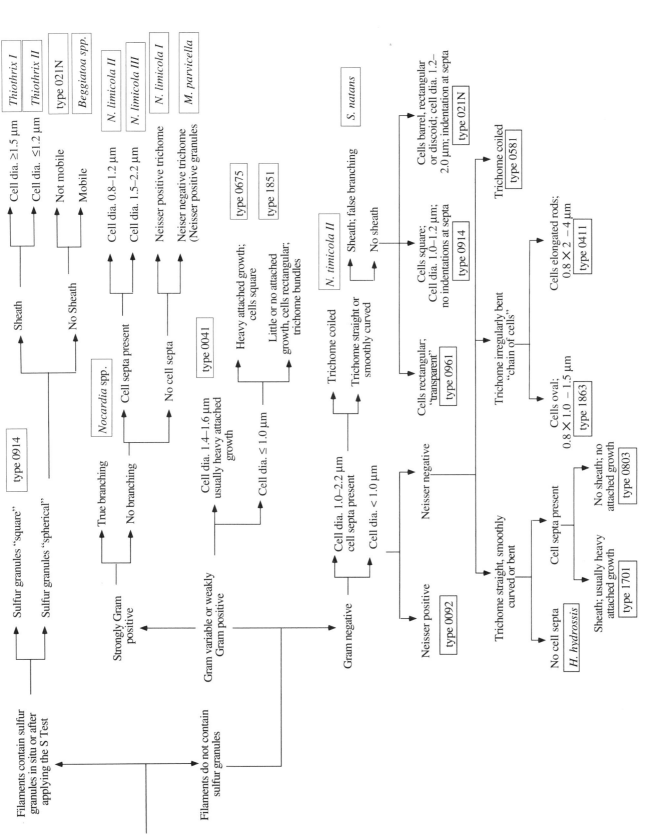

Figure 21. Dichotomous key for filamentous organism "identification" in activated sludge.

Table 17. Summary of Typical Morphological and Staining Characteristics of Filamentous Organisms Commonly Observed in Activated Sludge

| Filament Type | DIRECT ILLUMINATION OBSERVATION | | | | | | PHASE CONTRAST OBSERVATION AT 1000X | | | | | | | | | Notes |
	Gram Stain	Neisser Stain Trichome	Neisser Stain Granules	Sulfur Granules in situ	Sulfur Granules S test	Other Cell Inclusions	Trichome Diameter, μm	Trichome Length, μm	Trichome Shape	Trichome Location	Cell Septa Clearly Observed	Indentations at Cell Septa	Sheath	Attached Growth	Cell Shape and Size, μm	
S. natans	-	-	-	-	-	PHB	1.0-1.4	>500	St	E	+	+	+	-	round-ended rods 1.4×2.0	false branching
type 1701	-	-	-	-	-	PHB	0.6-0.8	20-80	St,B	I,E	+	+	+	++	round-ended rods 0.8×1.2	cell septa hard to discern
type 0041	+,V	-	-,+	-	-	-	1.4-1.6	100-500	St,SC	I,E	+	-	+	++,-	squares 1.4×1.5-2.0	Neisser positive reaction occurs
type 0675	+,V	-	-,+	-	-	-	0.8-1.0	50-150	St,SC	I	+	-	+	++,-	squares 1.0×1.0	Neisser positive reaction occurs
type 021N	-	-	-,+	-,++		PHB	1.0-2.0	50->1000	St,SC	E	+	+	-	-	barrels, rectangles, discoid 1.0-2.0×1.5-2.0	rosettes, gonidia
Thiothrix I	-,+	-	-,+	+,-	+	PHB	1.4-2.5	100->500	St,SC	E	+	+	+	-	rectangles 2.0×3.0-5.0	rosettes, gonidia
Thiothrix II	-	-	-,+	+,-	+	PHB	0.8-1.4	50-200	St,SC	E	+	-	+	-	rectangles 1.0×1.5	rosettes, gonidia
type 0914	-,+	-	-,+	-,+	-	PHB	1.0	50-200	St	E,F	+	-	-	-	squares 1.0×1.0	sulfur granules "square"
Beggiatoa spp.	-,+	-	-,+	+,-	+	PHB	1.2-3.0	100->500	St	F	-,+	-	-	-	rectangles 2.0×6.0	motile: flexing and gliding
type 1851	+ weak	-	-	-	-	-	0.8	100-300	St,SC	E	+,-	-	+	-,+	rectangles 0.8×1.5	trichome bundles
type 0803	-	-	-	-	-	-	0.8	50-150	St	E,F	+	-	-	-	rectangles 0.8×1.5	
type 0092	-	+	-	-	-	+	0.8-1.0	20-60	St,B	I	+,-	-	-	-	rectangles 0.8×1.5	
type 0961	-	-	-	-	-	-	0.8-1.2	40-80	St	E	+	-	-	-	rectangles 0.8-1.4×2.0-4.0	"transparent"
M. parvicella	+	-	+	-	-	PHB	0.8	50-200	C	I	-	-	-	-		large "patches"
Nocardia spp.	+	-	+	-	-	PHB	1.0	5-30	I	I	+,-	-	-	-	variable 1.0×1.0-2.0	true branching
N. limicola I	+	+	-	-	-	-	0.8	100	C	I,E	-	-	-	-		
N. limicola II	-,+	+,-	-	-	-	PHB	1.2-1.4	100-200	C	I,E	+	-	-	-	discs, ovals 1.2×1.4	incidental branching; Gram and Neisser variable
N. limicola III	+	+	-	-	-	PHB	2.0	200-300	C	I,E	+	+	-	-	discs, ovals 2.0×1.5	
H. hydrossis	-	-	-	-	-	-	0.5	10-100	St,B	E,F	-	-	+	-,+		"rigidly straight"
type 0581	-	-	-	-	-	-	0.5-0.8	100-200	C	I	-	-	-	-		
type 1863	-	-	-,+	-	-	-	0.8	20-50	B,I	E,F	+	+	-	-	oval rods 0.8×1.0-1.5	"chain of cells"
type 0411	-	-	-	-	-	-	0.8	50-150	B,I	E	+	+	-	-	elongated rods 0.8-2.0-4.0	"chain of cells"

LEGEND: + positive
    - negative
    +,- or -,+ variable, the first being most observed
    V variable, the first being most observed
    single symbol invariant

Trichome Shape
St Straight
B Bent
SC Smoothly curved
C Coiled
I Irregularly shaped

Trichome Location
E Extends from floc surface
I Forms mostly within the floc
F Free in liquid between the flocs

FILAMENTOUS ORGANISM DESCRIPTIONS

The short descriptions of each filamentous organism commonly observed in activated sludge presented here are based on information given by Eikelboom and van Buijsen (1981), as modified by our experience with filamentous organism characteristics in activated sludges from all over the world.

1. *Sphaerotilus natans*. (Figures 22a, b, and c; see also Figures 5a, 15d, 18a, and 19f). Relatively long (100–1000 μm) straight or smoothly-curved filaments composed of round-ended, rod-shaped cells ("sausage-shaped") (1.0 to 1.8 × 1.5 to 3.0 μm) contained in a clear, tightly-fitting sheath. Cell septa are clear with indentations at septa. Filaments radiate outward from the floc surface into the bulk solution. False branching frequently is observed, giving a "tree branch"-like appearance. Gram negative, Neisser negative, no sulfur granules; spherical PHB granules are frequently observed, often three per cell (Figure 22c). Cell shape can be rectangular when the cells are tightly packed within the sheath. An exocellular slime coating may occur under nutrient-deficient conditions. Attached epiphytic growth is uncommon, but may occur when the organism is growing slowly.

2. **type 1701.** (Figures 22d, e, and f; see also Figures 18b and f). Relatively short (20–100 μm), curved or bent filaments composed of round-ended, rod-shaped cells ("sausage-shaped") (0.7 to 1.0 × 1.0 to 2.0 μm) contained in a clear, tightly-fitting sheath. Cell septa are clear with indentations at septa. Filaments are found predominantly intertwined within the floc interior with only short filaments extending into the bulk solution. Occasionally filaments are observed growing out from the floc surface into the bulk solution and without attached growth; this indicates rapid growth, since the trichome is growing too fast to become covered with attached bacteria. No branching occurs. Gram negative, Neisser negative, no sulfur granules; spherical PHB granules are frequently observed. Significant amounts of attached growth of epiphytic bacteria are almost always observed, making observation of individual cells difficult.

3. **type 0041.** (Figures 23a, b, and c; see also Figures 17a and 18e, Plate 1c and Plate 2b). Straight or smoothly-curved filaments, 100–500 μm in length, composed of square-shaped cells (1.2 to 1.6 × 1.5 to 2.5 μm) contained in a clear, tightly-fitting sheath. A large form of type 0041, with a trichome width of 2.5–4.0 μ, is occasionally observed. In domestic wastewater activated sludge systems, type 0041 is most often observed inside the floc and covered with heavy, attached growth. In industrial wastewater activated sludge systems, it may occur extending from the floc surface or free in the bulk solution, and may not have any attached growth. Gram positive or Gram variable, tending to Gram positive when found within the floc and Gram negative when extending into the bulk solution. Neisser negative; Neisser-positive granules are observed infrequently; a Neisser-positive (light purple color) slime coating may be observed in some industrial wastewater activated sludges (Plate 2b). Because the staining characteristics are so variable it is best to rely more heavily on morphological features for identifying this organism. Intracellular granules rarely are observed. No sulfur granules are present. The sheath is difficult to detect and is best observed when cells are missing, particularly at the trichome apex (Figure 18e).

4. **type 0675.** (Figures 23d, e, and f). Very similar to type 0041, but smaller in trichome length (50–150 μm) and cell dimensions (0.8 to 1.0 μm). Covered with heavy, attached growth in domestic wastewater activated sludge; may lack attached growth in some industrial wastewater activated sludges. A sheath is present. Gram positive to Gram variable, Neisser negative. Neisser-positive granules occur, and no sulfur granules.

5. **type 021N.** (Figures 24a-f; see also Figures 14f, 16b, 19d and e, 20e, and Plate 1a). Filaments typically are 1.0 to 2.0 μm in width, 100 to >1000 μm in length, and taper from a thicker basal region, often exhibiting an inconspicuous holdfast, to a thinner apical region, often terminating in loosely-attached gonidia. Trichomes are straight, smoothly-curved or sometimes slightly coiled, and are found extending from the floc surface. (On occasion they have even been seen with a knot in the trichome!) Cell shape ranges from ovoid, rectangular or

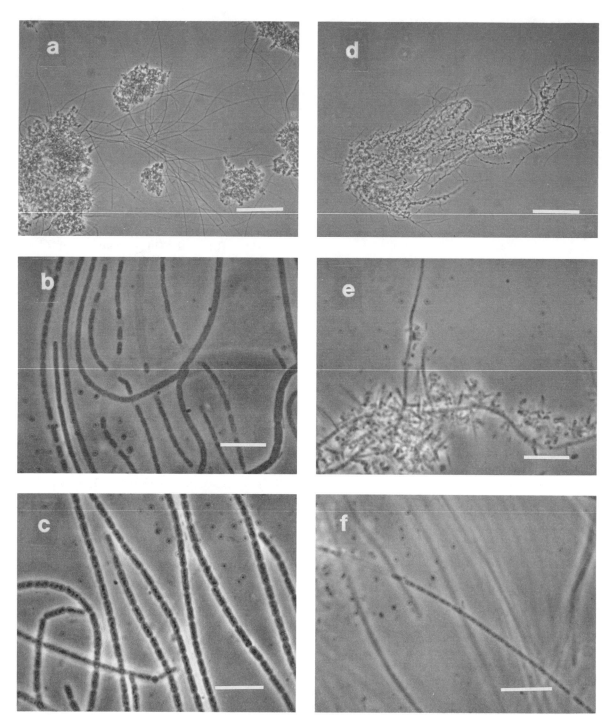

Figure 22. *Sphaerotilus natans*: *a.*, *b.* and *c.*; type 1701: *d.*, *e.* and *f.* (*a.* and *d.* 1000X phase contrast; bar = 100 μm; *b.*, *c.*, *e.* and *f.* 1000X phase contrast; bar = 10 μm).

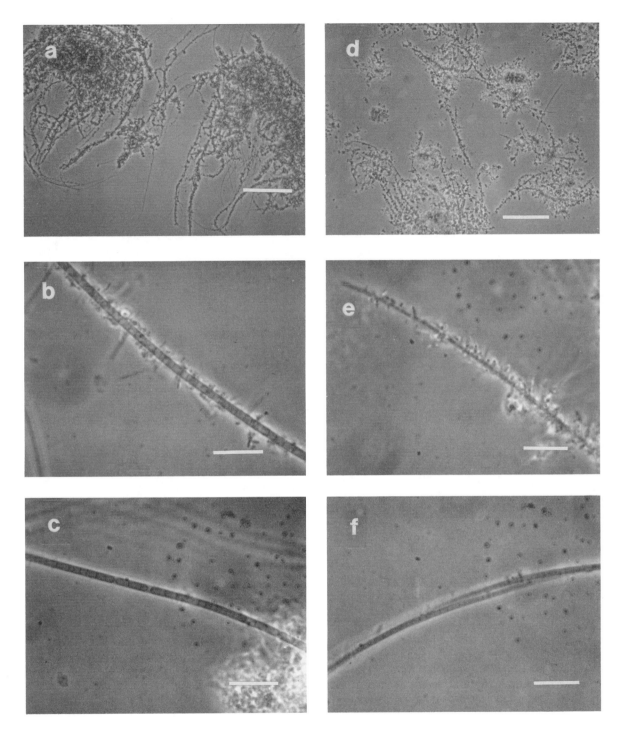

Figure 23. Type 0041: *a.*, *b.*, and *c.*; type 0675: *d.*, *e.* and *f.* (*a.* and *d.* 100X phase contrast; bar = 100 μm; *b.*, *c.*, *e.* and *f.* 1000X phase contrast; bar = 10 μm).

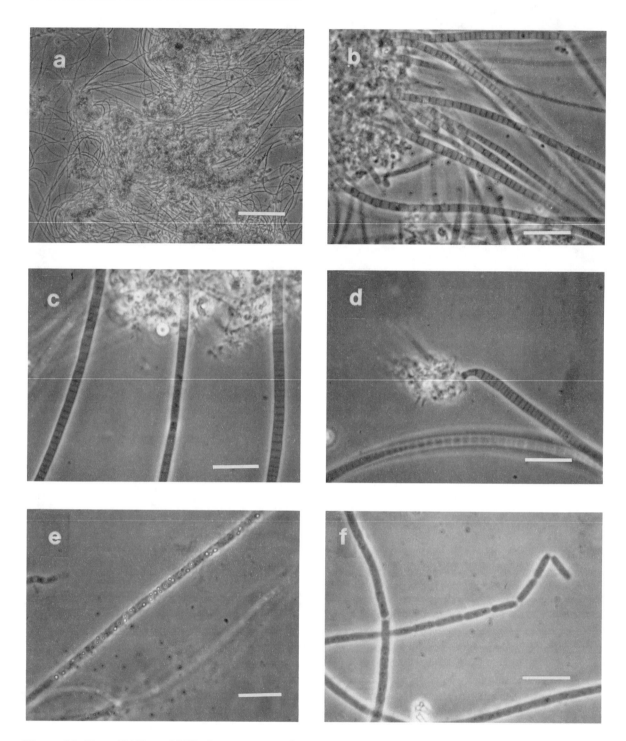

Figure 24. Type 021N: *a.* 100X phase contrast; bar = 100 μm; *b.−f.* 1000X phase contrast; bar = 10 μm.

barrel-shaped (often in the same filament) and always with well-defined cell septa and indentations at the septa. Cell sizes are generally in the 1–2 × 1–2 μm range. Gram negative and Neisser negative, and may contain Neisser-positive granules. Cells may stain slightly Gram positive when they contain sulfur granules. Spherical intracellular sulfur granules are observed in situ infrequently; response to the S test is often, but not always positive. Rosettes are observed infrequently. No attached growth occurs. No sheath is present; however, a heavy cell wall (still showing cross septa) often remains after cell lysis.

6. *Thiothrix I.* (Figure 25a-f; see also Figure 18d and 20c). Straight or smoothly-curved trichomes, 1.4 to 2.5 μm in width and 100–500 μm in length, found extending from the floc surface. Cells are rectangular (1.4 to 2.5 × 3–5 μm) with clear cell septa and without indentations at septa. No attached growth occurs. A prominent and heavy sheath is present. Gram negative and Neisser negative; however, a Gram positive reaction may occur when sulfur granules are present. Sometimes a Gram positive reaction occurs when a heavy sheath makes the decolorization step ineffective and incomplete stain removal is obtained. Neisser-positive granules may occur. Cells frequently contain sulfur granules in situ, and this organism responds strongly to the S test. Apical gonidia commonly are observed, and an inconspicuous holdfast can be present. Rosettes are observed infrequently.

7. *Thiothrix II.* (Figure 26a-f; see also Figures 18c and 20d). Straight or smoothly-curved filaments, 0.7 to 1.4 μm in width and 50–200 μm in length, found extending from the floc surface. No attached growth occurs. Cell septa without indentations are present. Cells are rectangular (0.7 to 1.4 × 1–2 μm). Gram negative and Neisser negative, with Neisser-positive granules sometimes present. Cells frequently contain spherical sulfur granules in situ, and this organism responds to the S test. Apical gonidia and rosettes are commonly observed. Trichomes may taper somewhat from base to tip. A sheath is present, but it is difficult to detect. Small amounts of attached growth may occur when the filament is not growing.

8. **type 0914.** (Figures 27a, b, and c). Straight or smoothly-curved filaments, 0.7 to 1.0 μm in width and 50–200 μm in length, found both extending from the floc surface and free in the bulk solution. May have incidental attached growth. Cells are square-shaped (1.0 × 1.0 μm) without constrictions at septa. No sheath is present. Gram negative and Neisser negative, but may stain Gram positive when substantial sulfur granules are present. Neisser positive granules may occur. May contain intracellular sulfur granules which appear square or rectangular. Does not respond to the S test.

9. *Beggiatoa* **spp.** (Figures 28a and b). Large, straight filaments, 1.0 to 3.0 μm in width and 100–500 μm in length, found free in the bulk solution and actively motile by gliding and flexing. Generally contains substantial spherical sulfur granules; cell septa may not be visible when sulfur granules are present. In the absence of sulfur granules the cells are rectangular (1–3 × 4–8 μm). No attached growth occurs. No sheath exists. Gram negative and Neisser negative; however, cells may stain Gram positive when substantial sulfur granules are present. Neisser positive granules may occur.

10. **type 1851.** (Figure 28c and d; see also Plate 1d). Straight or smoothly-curved trichromes, 0.8 to 1.0 μm in width and 100–300 μm in length, observed extending from the floc surface, or more commonly in bundles of intertwined filaments. A sheath is present, but difficult to observe. Cells are rectangular (0.8 × 1.5 to 2.5 μm) without indentations at septa. The septa are sometimes difficult to observe. Attached growth (often sparse) occurs, which is distinctly perpendicular to the trichrome surface. The Gram stain is weakly positive (a Gram positive "beaded" effect) or negative and the Neisser stain is negative. No sulfur granules occur.

11. **type 0803.** (Figure 28e). Straight or smoothly-curved filaments of uniform diameter (0.8 μm) and 50–150 μm in length, found extending from the floc surface or sometimes free in the bulk solution (especially in industrial wastewater activated sludges). No sheath and/or attached growth is observed. Cells are rectangular (0.8 × 1.5 to 2.0 μm) without constrictions at septa. The Gram and Neisser reactions

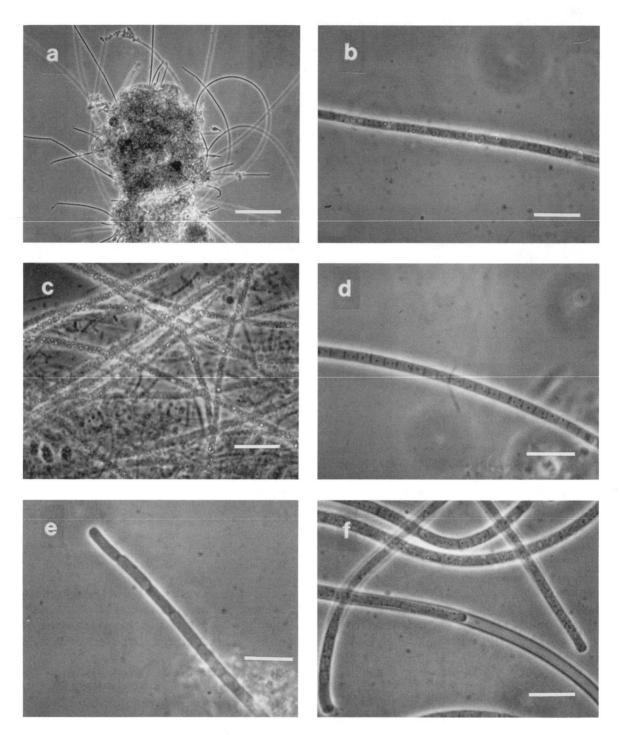

Figure 25. *Thiothrix* I: *a*. 100X phase contrast; bar = 100 μm; *b.*—*f*. 1000X phase contrast; bar = 10 μm.

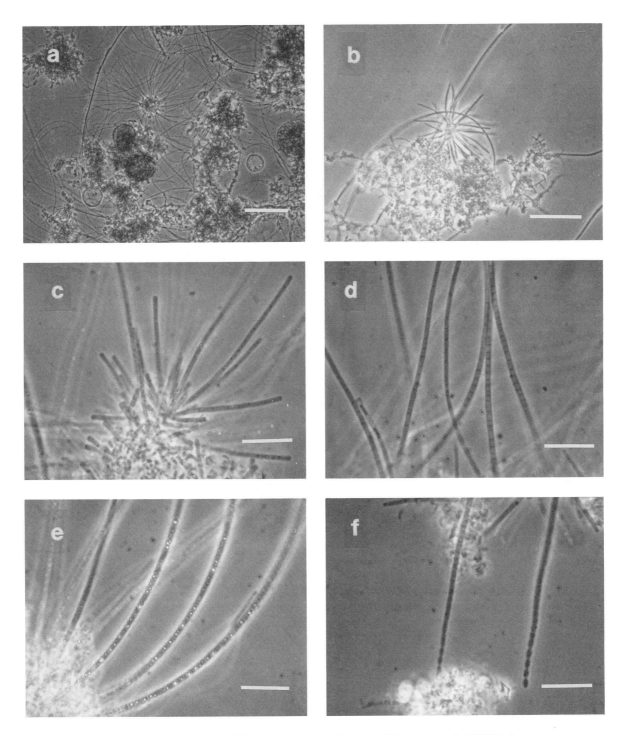

Figure 26. *Thiothrix* II: *a*. and *b*. 100X phase contrast; bar = 100 μm; *c.−f.* 1000X phase contrast; bar = 10 μm.

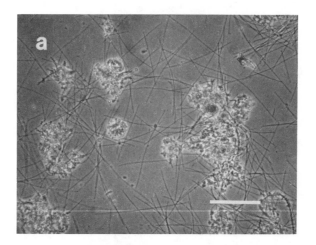

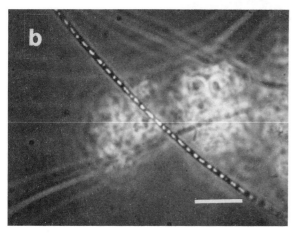
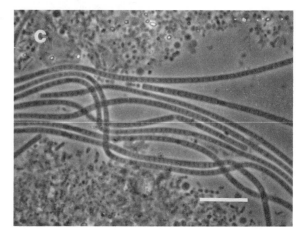

Figure 27. Type 0914: *a.* 100X phase contrast; bar = 100 μm; *b.* and *c.* 1000X phase contrast; bar = 10 μm (note: sulfur granules in *b.*).

are both negative and no sulfur granules occur.

12. **type 0092.** (Figure 28f; see also Plates 1b and 2c). Straight, irregularly-curved or bent filaments, 0.8 to 1.0 μm in diameter and 10–60 μm in length, found mostly within the floc. Cells are rectangular (0.8 × 1.5 μm) without constrictions at septa, or at times these are hard to observe. Neither attached growth nor a sheath are present. Cells stain Gram negative and the entire trichome stains Neisser positive (purple). No sulfur granules occur. This filament often is overlooked, or its abundance is underestimated until the Neisser stained slide is examined. Filaments appear wider (1.0 to 1.2 μm) in dried, Neisser and Gram stained smears than when observed in wet mounts.

13. **type 0961.** (Figures 29a and b). Straight trichomes, 0.8 to 1.4 μm in diameter and 20–150 μm in length, observed extending from the floc surface. No attached growth occurs. The cells are rectangular (0.8 to 1.4 × 2–4 μm). A true sheath is not present; however, a slime coating may be present which can be seen best as an empty "cuff" at the trichome apex. The filament is Gram negative and Neisser negative, and no sulfur granules occur. The cells appear "transparent" without any internal structures.

14. *Microthrix parvicella.* (Figures 29c, d, and e; see also Figure 16e and Plates 1e and 2e). Irregularly-coiled filaments, 0.6 to 0.8 μm in diameter and 50–200 μm in length, found in tangles in the floc or as loose "patches" free in the bulk solution. Neither attached growth nor a sheath are present. There is no branching. Cell septa are not observed; however, substantial intracellular granules may occur which give a "beaded" effect.

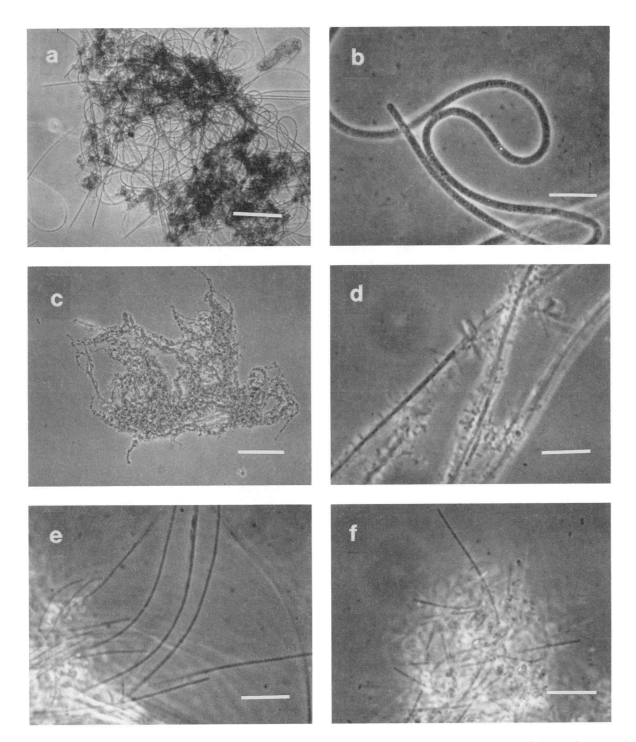

Figure 28. *Beggiatoa* spp: *a.* and *b.*; type 1851: *c.* and *d.*; type 0803: *e.*; and type 0092: *f.* (*a.* and *c.* 100X phase contrast; bar = 100 μm; *b.*, *d.*, *e.* and *f.* 1000X phase contrast; bar = 10 μm).

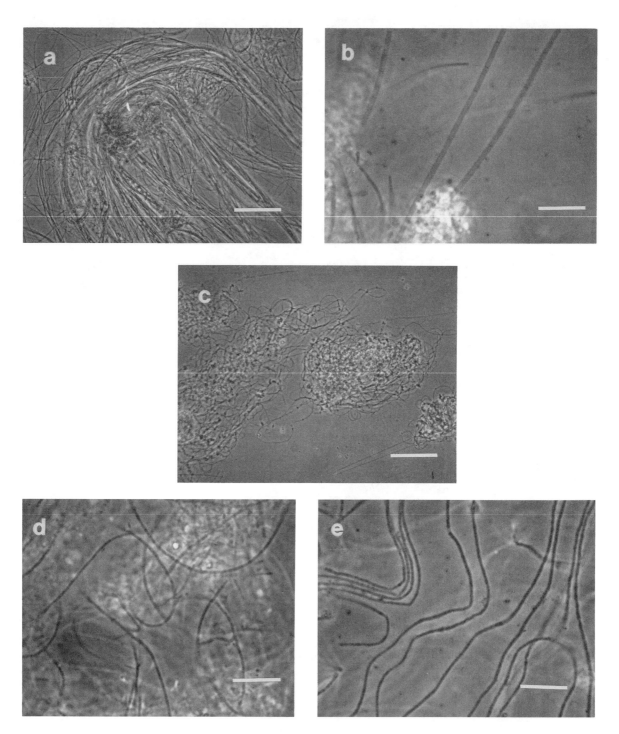

Figure 29. Type 0961: *a.* and *b.*; and *Microthrix parvicella*: *c.*, *d.* and *e.* (*a.* and *c.* 100X phase contrast; bar = 100 μm; *b.*, *d.* and *e.* 1000X phase contrast; bar = 10 μm).

The filament stains strongly Gram positive and Neisser negative, commonly with Neisser positive granules. Short, clear spaces may occur in the filament that should not be confused with a sheath.

15. *Nocardia* **spp.** (Figures 30a, b, and c; see also Figures 15b and 16f, and Plates 1f and 2f). Irregularly-bent, short filaments, 1.0 μm in diameter and 5–30 μm in length, found mostly within the floc but also free in the bulk solution, especially when foam is trapped in an aeration basin. A branched (true branching) mycelium often is observed. No sheath and no attached growth occur. Cell shape is somewhat irregular (1.0 × 1.0 to 2.0 μm), and septa without constrictions are clearly visible. The filament is Gram positive and Neisser negative, and Neisser positive granules commonly are observed. No sulfur granules are present. PHB granules commonly are observed. The abundance of this organism is best assessed from the Gram stained preparation.

16. *Nostocoida limicola I.* Bent and irregularly-coiled filaments, 0.6 to 0.8 μm in diameter and 100–200 μm in length, found within the flocs and free in the bulk solution. Cell septa are hard to observe; when observed, the cells are oval (0.6 to 0.8 μm diameter). No sheath and no attached growth occur. The organism is Gram positive and Neisser positive. No sulfur granules occur. No attached growth. This organism resembles *M. parvicella*, except in its Neisser-staining properties.

17. *Nostocoida limicola II.* (Figures 30d, e, and f; see also Figure 19c and Plate 2d). Bent and irregularly-coiled filaments, 1.2 to 1.4 μm in diameter and 100–200 μm in length, found mostly within the floc. Cell septa are clear, with oval cells (1.2 to 1.4 μm diameter) and indentations at septa. No sheath and no sulfur granules are present. PHB granules are commonly observed. Gram and Neisser staining reactions are variable. Most often observed is a Gram negative, but a Gram positive reaction can occur. Most often the entire trichome stains Neisser positive (purple), but it can be Neisser negative at times. Incidental branching is observed. *Note:* Eikelboom and van Buijsen (1981) state that *N. limicola II* is both Gram positive and Neisser positive, and that a bacterium closely resembling *N. limi-*

cola II occurs in some industrial wastewater activated sludge systems, differing from *N. limicola II* by being Gram negative and Neisser negative. Both forms are considered *N. limicola II* here.

18. *Nostocoida limicola III.* Bent and irregularly-coiled filaments, 1.6 to 2.0 μm in diameter and 200–300 μm in length, found mostly extending from the floc. No attached growth and no sheath occur. Cell septa are clear, with oval cells (1.6 to 2.0 μm diameter) and indentations at cell septa. PHB granules commonly are observed. Usually the Gram and Neisser reactions are both positive but sometimes they can be negative. No sulfur granules are observed.

19. *Haliscomenobacter hydrossis.* (Figures 31a, b, and c). Straight or bent, thin filaments, 0.5 μm in diameter and 10–100 μm in length, found radiating outward from the floc surface or free in the bulk solution. A sheath is present. No cell septa are observed; however, empty spaces in the trichome commonly are observed. The Gram and Neisser stains are both negative. No sulfur granules are observed. Filaments may occur in bundles, and attached growth is variable, ranging from rare to abundant. This filament can be easily overlooked at 100X observation, especially with direct illumination.

20. **type 0581.** (Figure 31d). Smoothly-coiled filaments, 0.4 to 0.7 μm in diameter and 100–200 μm in length, found mostly within the floc but may occur in "patches" free in solution. No sheath and no cell septa are observed. There are no sulfur granules. No attached growth exists. The Gram and Neisser stains are both negative. This filamentous organism appears similar to *M. parvicella*, but differs in its Gram and Neisser staining reactions.

21. **type 1863.** (Figure 31e). Short, irregularly-bent filaments, 0.8 μm in diameter and 50 μm in length, found extending from the floc surface and free in the bulk solution. No sheath and no attached growth occur. Cells are oval-shaped rods (0.8 × 1.5 μm), and appear as a "chain of cells" with indentations at the septa. No rigid trichome exists. The Gram and Neisser stains are both negative; Neisser positive granules may occur. No sulfur granules are observed.

22. **type 0411.** (Figure 31f). Irregularly-bent trichomes, 0.8 μm in diameter and 50–150 μm

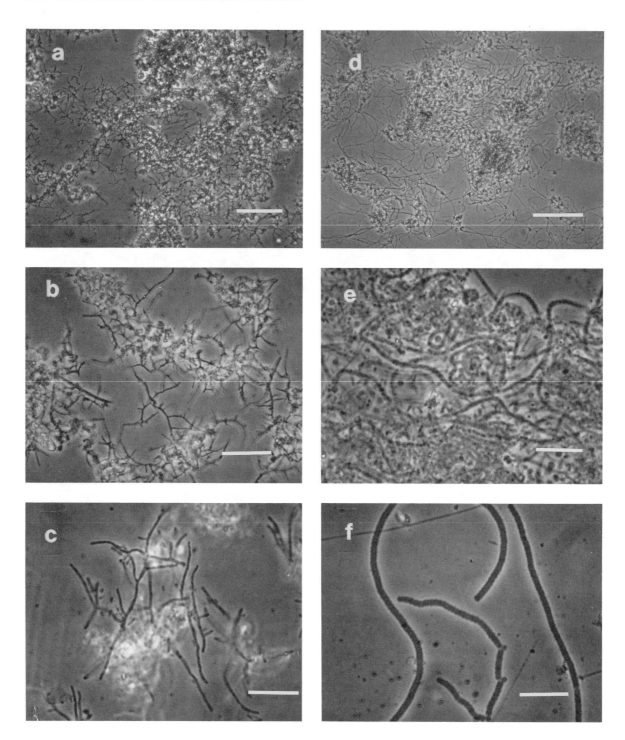

Figure 30. *Nocardia* spp: *a.*, *b.* and *c.*; and *Nostocoida limicola* II: *d.*, *e.* and *f.* (*a.* and *d.* 100X phase contrast; bar = 100 μm; *b.* 400X phase contrast; bar = 25 μm; *c.*, *e.* and *f.* 1000X phase contrast; bar = 10 μm).

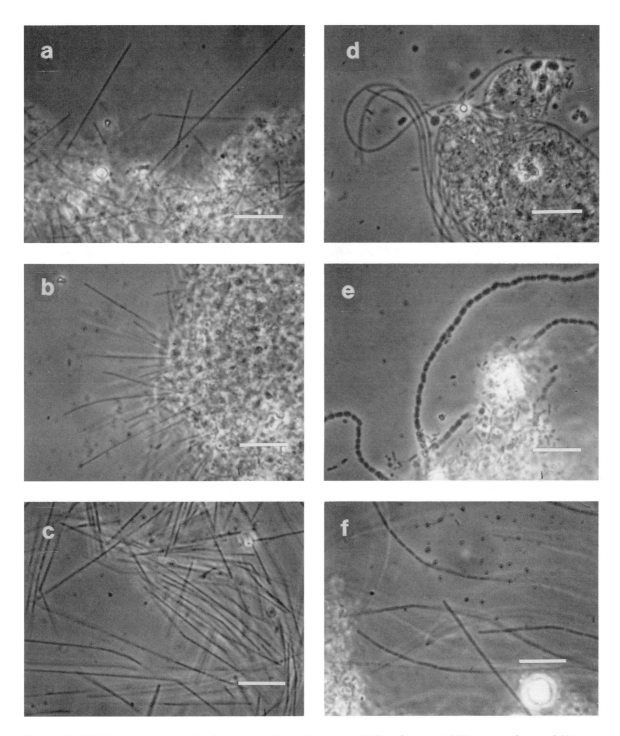

Figure 31. *Haliscomenobacter hydrossis*: *a.*, *b.* and *c.*; type 0581: *d.*; type 1863: *e.*; and type 0411; *f.* (all 1000X phase contrast; bar = 10 μm).

in length, found extending from the floc surface. Filaments composed of elongated, rod-shaped cells (0.8 × 2–4 μm); constrictions at septa give the appearance of a "chain of cells." No sheath and no attached growth are seen. The Gram and Neisser stains are both negative. No sulfur granules are observed.

23. **type 1702.** (Figure 32a). Short, straight or bent filaments, 0.6 to 0.7 μm in diameter and 20–80 μm in length, found within the floc and extending from the floc surface. Cell septa are absent and a sheath is present. Incidental attached growth occurs. The Gram and Neisser stains are both negative. No sulfur granules occur.

24. **type 1852.** (Figure 32b). Straight or slightly bent filaments, 0.6 to 0.8 μm in diameter and 20–80 μm in length, found extending from the floc surface. Cells are rectangular (0.6 to 0.8 × 1.0 to 2.0 μm) without constrictions at septa. No sheath and no attached growth are observed. The Gram and Neisser stains are both negative. No sulfur granules occur. This filament appears "transparent" similar to type 0961.

25. **type 0211.** (Figure 32c). Bent and twisted filaments, 0.3 to 0.5 μm in diameter and 20–100 μm in length, found extending from the floc surface. Cells are rod-shaped with clear constrictions at cell septa. No sheath and no attached growth occur. The Gram and Neisser stains are both negative. No sulfur granules occur.

26. *Flexibacter* **spp.** (Figure 32d). Short, straight, or smoothly-curved filaments, 1.0 μm in diameter and 20–40 μm in length, found free in the bulk solution. This organism is motile by slow gliding and flexing. No sheath and no attached growth occur. Cell septa may be lacking. The Gram and Neisser stains are both negative. PHB granules are commonly observed.

27. *Bacillus* **spp.** (Figure 32e). Rounded-rods in irregularly-shaped chains of cells, 0.8 to 1.0 μm in diameter and 20–50 μm in length, found mostly at the edges of the floc. The Gram and Neisser stains are both negative. No sheath and no sulfur granules are observed.

28. *Cyanophyceae.* (Figure 32f). Straight, large filaments, 2.0 to 5.0 μm in diameter and 100–500 μm in length, found free in the bulk solution. Cells are square to rectangular (2–5 × 2–8 μm) with clear septa. No

sheath and no attached growth are observed. The organisms are often motile by slow gliding. The filaments have a distinct green color when viewed under direct illumination. The Gram stain usually is negative, but sometimes a slight Gram positive reaction occurs. The Neisser stain is negative. No sulfur granules occur.

29. *fungi.* (Figures 33a, b, and c; see also Figure 15a). Very large trichomes, 3–8 μm in diameter and 300–1000 μm in length, found mostly in the floc. Cells are rectangular (3–8 × 5–15 μm) and contain intracellular granules and organelles; cytoplasmic streaming may be observed. True branching occurs. The Gram and Neisser stains are both negative. No sulfur granules and no sheath occur, although a heavy cell wall is present.

RESULTS OF FILAMENTOUS ORGANISM IDENTIFICATION IN ACTIVATED SLUDGE

The types of filamentous organisms observed in activated sludge have been investigated in a systematic fashion by several workers including Cyrus and Sladka (1970), Hunerberg et al., (1970), Farquhar and Boyle (1971a; 1971b), Sladka and Ottova (1973), Eikelboom (1975; 1977), Eikelboom and van Buijsen (1981), Richard et al., (1982), Wagner (1982), Strom and Jenkins (1984), Blackbeard et al., (1986), and Richard (1989). These studies have established that 25 to 30 different filamentous organisms can be found in activated sludge.

Surveys have been conducted in the USA (Richard et al., 1982; Strom and Jenkins, 1984; Richard, 1989) (Table 18), in Northern Europe (Eikelboom, 1977; Wagner, 1982), and in South Africa (Blackbeard et al., 1986) (Table 19), using Eikelboom's filamentous organism identification procedures and key, to examine the relative occurrence of various filamentous organism types in bulking and foaming activated sludge. These surveys have demonstrated that the same filamentous organism types are observed ubiquitously in activated sludge and that approximately 10–12 types account for the great majority of all bulking and foaming episodes. The relative frequency of occurrence of dominant individual filamentous organisms in bulking sludges can vary significantly between geographical areas (Table 19). These differences can be explained in

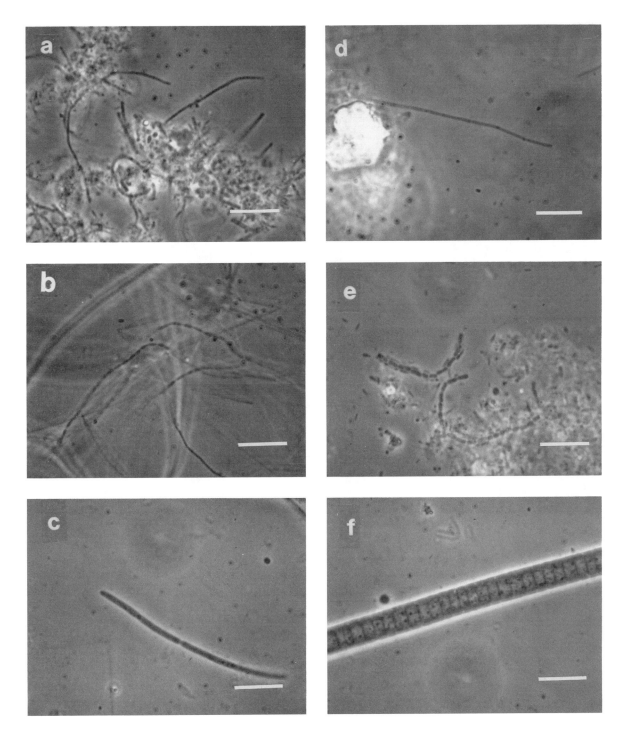

Figure 32. Type 1702: *a.*; type 1852: *d.*; type 0211: *b.*; *Flexibacter* spp: *c.*; *Bacillus* spp: *e.*; and a blue-green bacterium (cyanophyceae); *f.* (all 1000X phase contrast; bar = 10 μm).

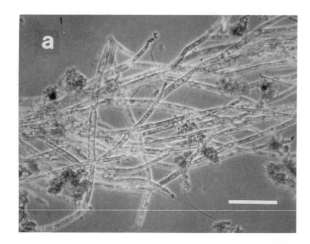
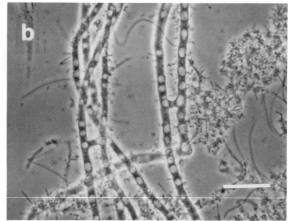
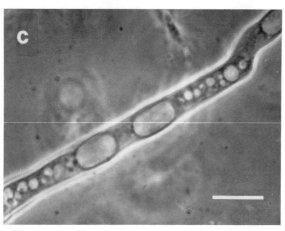

Figure 33. Fungus: *a*. 100X phase contrast; bar = 100 μm; *b*. 400X phase contrast; bar = 25 μm; *c*. 1000X phase contrast; bar = 10 μm.

terms of differences in wastewater type and strength; in plant operating conditions [particularly F/M (or MCRT)], DO concentration, and temperature; and aeration basin design, especially the initial mixing conditions between the influent wastewater and the RAS and the presence of initial anaerobic, anoxic, or otherwise unaerated zones in the aeration basin.

DIAGNOSIS OF CAUSES OF SOLIDS SEPARATION PROBLEMS FROM MICROSCOPIC EXAMINATION OF ACTIVATED SLUDGE

General

Much can be learned about how an activated sludge will behave during solids separation processes from a careful microscopic examination of

it, especially if the observer is "aware" and keeps an open mind—you can find some strange things in activated sludge!! Such an examination should always be conducted when addressing solids separation problems. These data are completely independent of any other analysis done for resolving such problems. Furthermore, a microscopic observation of the flocs and filamentous organisms provides a measurement that reflects not just the current conditions, but to some degree the conditions that have existed over at least the previous one or two MCRTs. Microscopic observation of activated sludge has its greatest value when used in conjunction with other data such as chemical analyses, operating records, and plant design data.

In this section we will go into great detail on the information that can be gleaned from determining the types of filamentous organisms present. While this is very useful in problem diag-

Table 18. Filament Abundance in USA Bulking and Foaming Activated Sludges[a]

Rank	Filamentous Organism	Percentage of Treatment Plants with Bulking or Foaming Where Filament was Observed to be:	
		Dominant	Secondary
1	*Nocardia* spp.	31	17
2	type 1701	29	24
3	type 021N	19	15
4	type 0041	16	47
5	*Thiothrix* spp.	12	20
6	*Sphaerotilus natans*	12	19
7	*Microthrix parvicella*	10	3
8	type 0092	9	4
9	*Haliscomenobacter hydrossis*	9	45
10	type 0675	7	16
11	type 0803	6	9
12	*Nostocoida limicola* (types I, II and III)	6	18
13	type 1851	6	2
14	type 0961	4	6
15	type 0581	3	1
16	*Beggiatoa* spp.	1	4
17	fungi	1	2
18	type 0914	1	1
–	all others	1	–

[a]Combined results of Richard et al., 1982 and Strom and Jenkins, 1984; 525 samples from 270 treatment plants.

nosis, it is not the only thing that can be learned from a microscopic examination of activated sludge. Much useful information can be obtained from observing floc characteristics and the presence of other types of organisms or nonbiological particles.

Nonbiological Particles

The nonbiological particles present in activated sludge flocs can be informative (Figure 34). Most activated sludges contain a few paper fibers, most likely from toilet tissue. Pulp and paper wastewater activated sludges always contain fibers derived from wood. The observation of many paper fibers (Figure 34a) in municipal wastewater activated sludge suggests that the plant does not have primary clarifiers, because most paper fibers are settled out in primary clarifiers. Also seen in the activated sludge from this type of plant are hairs (Figure 34c), plant ducts (Figure 34d), and grease particles (Figure 34e).

The presence of large amounts of amorphous organic particles in a municipal wastewater activated sludge suggests the presence of recycled digested sludge solids from somewhere in the solids handling processes. In severe cases of such solids recycle, the odor of digested sludge will be apparent in the activated sludge sample.

Often this observation can be coupled with the finding of low percent VSS in the mixed liquor and the presence of a turbid secondary effluent.

Hydrogen sulfide reacts rapidly with iron (which is present in most wastewaters) to form a black precipitate of iron sulfide which accumulates in the floc (Plate 3c). The finding of significant iron sulfide in the floc is an indication of waste septicity or high hydrogen sulfide somewhere in the system. Often, this is a sign of septic return flows to the aeration basin, usually from a solids handling process.

When iron salts are added to the primary effluent or the mixed liquor (for example, for phosphate removal by simultaneous precipitation) it is usually possible to see yellow amorphous areas in the flocs consisting of the ferric phosphate/hydroxide precipitates (Plate 3d). Note that, since phase contrast illumination distorts the true colors of samples, it is necessary to confirm the true color of the sample by looking at it under direct

Table 19. Comparison of Dominant Filamentous Organisms in Bulking and Foaming Activated Sludge from Several Geographical Areas

Filamentous Organism	USA[a]	Netherlands[b,e]	Germany[c,e]	South Africa[d]	Colorado, USA[f]
Nocardia spp.	1	—	—	7	2
type 1701	2	5	8	—	1
type 021N	3	2	1	—	10
type 0041	4	6	3	6	7
Thiothrix spp.	5	19	—	—	5
Sphaerotilus natans	6	7	4	—	8
Microthrix parvicella	7	1	2	3	2
type 0092	8	4	—	1	8
Haliscomenobacter hydrossis	9	3	6	—	—
type 0675	10	—	—	5	5
type 0803	11	9	10	8	—
Nostocoida limicola (types I, II and III)	12	11	7	—	—
type 1851	13	12	—	4	4
type 0961	14	10	9	—	—
type 0581	15	8	—	—	—
Beggiatoa spp.	16	18	—	—	—
fungi	17	15	—	—	—
type 0914	18	—	—	2	—

[a]Richard et al., 1982 and Strom and Jenkins, 1984: 525 samples from 270 treatment plants.
[b]Eikelboom, 1977: 1100 samples from 200 treatment plants.
[c]Wagner, 1982: 3500 samples from 315 treatment plants.
[d]Blackbeard and Ekama, 1984; Blackbeard et al., 1986; Blackbeard et al., 1986: 111 treatment plants including 26 BNR plants. Bulking and nonbulking sludges included.
[e]*Nocardia* spp. not included in these surveys since it was classified as a foaming organism rather than a bulking organism. The surveys were confined to bulking organisms.
[f]Richard, 1989: 24 domestic waste treatment plants monitored at least quarterly for one year.

illumination. The presence of inorganic precipitates in activated sludge also causes a reduction in its volatile fraction.

In the PACT process (Hutton and Robertaccio, 1975) powdered carbon is added to the activated sludge system. In such activated sludges it is easy to see the particles of activated carbon microscopically, even at a magnification of 100X (Figure 34b). The carbon particles are thick and tough enough to prevent one from preparing a very thin preparation by pressing on the cover slip, so that at a magnification of 1000X it is difficult to focus the microscope properly.

Many other nonbiological particles are found in activated sludges treating various industrial wastewaters. Over the years we have seen paper fibers, dye particles, resin beads, diatomaceous earth, catalyst support particles, oil droplets (Figure 34f), blue fibers from clothing manufacture, and starch granules.

ACTIVATED SLUDGE MICROBIOLOGICAL FEATURES OTHER THAN FILAMENTS

Microscopic observation of the condition of activated sludge flocs and the occurrence of microorganisms other than filaments can be used for judging the nature of the growth environment in the aeration basin. Important floc characteristics are floc strength or its absence (dispersed growth), bacterial diversity, the presence of specific microorganisms indicative of growth conditions, and the presence of algae, protozoa, and other higher life forms.

In a typical activated sludge floc developed on domestic wastewater, one usually observes a wide variety of bacterial morphological forms, indicative of the growth of a variety of bacterial species on a mixture of substrates. Activated sludge flocs

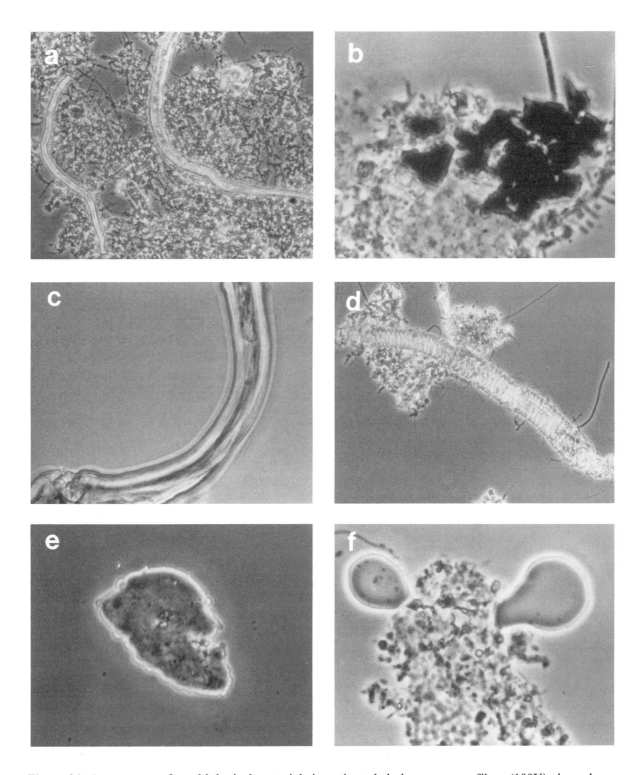

Figure 34. Appearance of nonbiological materials in activated sludge: *a*. paper fibers (100X); *b*. carbon particles (1000X); *c*. hair (200X); *d*. plant ducts (100X); *e*. grease (200X); and *f*. oil droplets (400X) (all phase contrast).

developed on some industrial wastewaters often exhibit a more limited variety of morphological types of bacteria, indicative of their growth on a waste with a simpler composition. Such activated sludges may not be as resilient to changes in conditions, either environmental or in wastewater type, as those developed on a more diverse substrate (such as domestic wastewater).

For an activated sludge to produce a well-clarified effluent, the level of individual bacterial cells and small aggregates of cells (both termed dispersed growth) must be low. Large amounts of dispersed growth (Figure 12b) in the aeration basin will produce a turbid effluent. Dispersed growth is encountered under a variety of conditions.

An activated sludge sample that has been stored in a full bottle, especially if it has become warm and septic, will develop dispersed growth. This is an artifact of improper sample collection and storage; it occurs more rapidly with low MCRT activated sludges than with high MCRT activated sludges. A fresh, properly collected and stored sample should always be examined.

High F/M (low MCRT) increases the amount of dispersed bacteria in an activated sludge. In municipal wastewater activated sludges, dispersed growth can start to have a noticeable impact on effluent quality at MCRTs of less than about 2 days. For activated sludges developed on high organic strength, readily degradable, soluble industrial wastewaters, dispersed growth can occur at much lower F/M values. Spills and shock loads in industrial wastewater activated sludge systems typically result in dispersed growth and turbid effluents.

Dispersed growth is common in biofilter effluents, especially from high rate biofilters, and is the cause of the turbidity often seen in these effluents. Dispersed growth arises because of the inadequate bioflocculation time afforded by high rate biofilters. The dispersed microorganisms can be removed by bioflocculating the biofilter effluent using processes such as the trickling filter/solids contact (TF/SC) process (Norris et al., 1982).

The two causes of dispersed growth discussed above result from inadequate bioflocculation (no floc formation). Dispersed growth can also result from the breakup or deflocculation of existing flocs. Three "generic" causes for this type of dispersed growth have been identified—mechanical shearing, the presence of surfactants, and the action of toxicants. The mechanical shearing or breakup of flocs occurs in activated sludge sys-

tems with vigorous methods of aeration (e.g., mechanical aerators and coarse bubble diffused air) and in "overoxidized" or "old" activated sludges. This type of floc is referred to as pin-floc or "pin-point" floc. In such sludges the majority of the biomass, with the exception of the small dispersed floc, usually settles rapidly (SVI < 75 mL/g).

Pin-floc can be rectified by reflocculation of the small flocs with the larger ones. This can be done by ensuring gentle transfer of mixed liquor between the aeration basin and the secondary clarifier and providing a gently mixed flocculation zone in the secondary clarifier. To achieve this in a circular secondary clarifier, one should provide a large tangentially-fed center well with sufficient hydraulic detention time to provide for reflocculation of the small flocs.

Slowly biodegradable surfactants (e.g., the branched chain alkyl phenol ethoxylate nonionic surfactants) can cause floc dispersion. This dispersion effect can also be noticed in the behavior of the secondary clarifier sludge blanket which, instead of possessing a distinct interface between the settled sludge and the supernatant liquid, becomes diffuse, showing a very gradual transition from the sludge layer to the supernatant over a distance of up to several feet. An analysis for surfactants (anionic and nonionic) and of surface tension should be made for confirmation. Surface tension values of below 60 dynes/cm^2 in the liquid from the aeration basin or in the secondary effluent are indicative of such a problem. Also, the presence of a white, frothy foam on the surface of the treatment unit should raise one's suspicion that surfactants may be present. We have seen this surfactant floc dispersion problem in activated sludges treating surfactant wastewater, pharmaceutical wastewater, textile wastewaters, commercial laundry wastewater, and on occasion in the wastewaters from industries that engage in periodic equipment cleaning, e.g., dairies, cheese manufacturers, canneries. This type of floc dispersion problem can vary seasonally, being worse when the wastewater is colder because the biodegradation rate of some of the causative surfactants is significantly decreased at low temperatures.

The presence of toxicants in the influent wastewater, especially heavy metals, can result in dispersed growth due to deflocculation. Filamentous organisms, if present in the activated sludge, are often the first microorganisms to be affected by toxic metals. The SVI decreases rather rapidly, and freely dispersed damaged filament trichomes can be observed microscopically in the supernatant above the settled sludge. If the toxicity is severe enough or reoccurring, deflocculation will occur. Often, protozoa, usually flagellates, will "bloom" following deflocculation, due to the sudden availability of a food source (the deflocculated bacteria). If the toxicity event is severe enough, these protozoa are killed and their lysed cell contents can cause a foam. Microscopic examination of this foam will reveal the presence of dead protozoans and protozoan fragments in it. The activated sludge BOD removal usually declines or ceases following this event. Activated sludge affected by severe heavy metal toxicity will have a much higher than normal heavy metal content in the MLSS. This can be determined by chemical analysis.

A similar sequence of events can occur during RAS chlorination for bulking control (see Chapter 3). If the initial mixing of chlorine and RAS is poor, or if the chlorine dose is too high, the floc breaks up, forming small floc fragments. Dispersed growth and a turbid effluent results.

High aeration basin temperature can cause dispersed growth. This usually occurs in industrial wastewater activated sludges systems at temperatures above approximately 35–40°C. A common observation is the occurrence of an episode of dispersed growth and high effluent turbidity as the aeration basin temperature increases from below about 35°C to above this value. The dispersed growth and effluent turbidity usually subside after a few days as a new thermotolerant group of floc bacteria develops. Dispersed growth episodes occur also as the temperature decreases through this range because the thermotolerant floc formers leave the system and are replaced by mesophilic floc formers. For this reason, in activated sludges systems operated at high temperatures (>35°C) it is important to limit temperature variations as much as possible.

Dispersed growth and a turbid effluent can be caused by the growth of several types of microorganisms free in suspension and not attracted to the flocs. These organisms can be filamentous (e.g., *H. hydrossis, N. limicola* and type 1863) or one of several types of single-celled bacteria of unique appearance. One of the more common of these is a Gram negative and Neisser positive staining group of four cells, termed the Neisser positive tetrad (Plate 3a). The occurrence of this organism and the turbidity problems associated with it have been related to both nutrient deficiency and low F/M operation.

A Neisser-positive staining bacterium that grows in grape-like clusters inside the activated sludge flocs occurs in systems with initial anoxic or anaerobic zones (selectors) in the aeration basin (Plate 3b). These cell clumps are thought to be Acinetobacters and the Neisser positive staining reaction is thought to be associated with internally stored inorganic polyphosphate granules that accumulate when these organisms are cycled through an anaerobic/aerobic environment. These types of Neisser-positive staining cells are found in significant amounts in activated sludge systems exhibiting EBPR (e.g., processes such as Bardenpho, A/O, Phostrip, and UCT).

The presence of large amounts of yeast (Figure 35a) in activated sludge is unusual and suggests one of four conditions: (a) the presence of yeast in the influent wastewater (e.g., wastewaters from bakery, brewery, or other fermentation process); (b) the existence of fermentative conditions somewhere in the collection system or the treatment process; (c) low pH <6.0 (similar to fungi, to which yeasts are related); and (d) severe phosphorus deficiency. We have seen yeasts proliferate in the activated sludge in a system aerated with fine pore diffusers when the diffusers became clogged and the oxygen supply was insufficient. Another yeast overgrowth incident was associated with the fermentation of a sugar wastewater and domestic wastewater mixture in the primary clarifiers of the plant. Septic conditions in raw sludge gravity thickeners can produce conditions that favor yeast growth. The yeasts are recycled back into the activated sludge in the gravity thickener overflow. "Gassing" of the primary clarifiers often will accompany the observation of high yeast concentrations in the wastewater influent.

The presence of zoogloeal colonies (either amorphous or fingered in shape) (Figures 12c and d) can be indicative of several conditions. Both types of zoogloeas are most often observed in high F/M systems, especially if the waste contains readily biodegradable, soluble organic compounds and the pH is low. For these reasons amorphous zoogloea often develop in selector system activated sludges. In coupled fixed-film/activated sludge systems with no intermediate clarifier, zoogloeas often wash into the activated sludge mixed liquor from the fixed-film system.

Flocs from activated sludge systems that incorporate properly functioning selectors often have unique features. Besides the already discussed presence of zoogloeas (usually the amorphous type), these flocs often appear to be tightly structured and "strong." Chudoba (1989) claims that

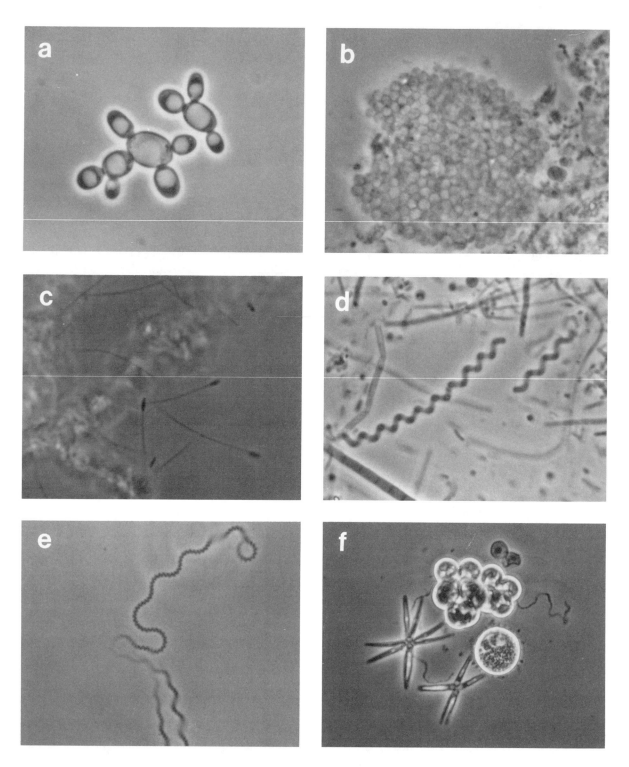

Figure 35. Microorganisms other than filaments of significance in activated sludge: *a*. yeast (1000X); *b*. nitrifying bacteria (1000X); *c*. *Hyphomicrobium* spp. (1000X); *d*. *Spirillum* spp.(1000X); *e*. *Spirochaete* (1000X); *f*. algae (400X) (all phase contrast).

these flocs gain their strength through the presence of a tightly fitting exocellular glycolax. In anoxic and anaerobic selector sludges, Neisser positive staining cell clusters (Plate 3b) and individual Neisser positive staining cells are seen. These selector sludges are not always devoid of filamentous organisms, but those present usually are quite short in length and often have substantial attached growth — a sign of slow growth. Often their staining reactions are not typical; all in all they appear stressed.

Nitrifying bacteria, when present in significant amounts in the activated sludge, can be observed microscopically at 1000X magnification as dense, rounded microcolonies of bacteria, generally found at the floc edge (Figure 35b). Most denitrifying bacteria are indistinguishable from other bacteria in the activated sludge floc; however, the denitrifying bacterium *Hyphomicrobium* spp. is easily identified due to its unique "bean on a stalk" morphology (Figure 35c). Observation of these two microorganisms has proved useful in industrial wastewater systems in judging ammonia overaddition and the accompanying nitrification-denitrification problems (e.g., floating sludge).

Influent wastewater or aeration basin septicity can be detected by the presence of high amounts of either *Spirillum* spp. or Spirochaetes (Figures 35d and e) found free in the fluid between the activated sludge flocs. These bacteria grow preferentially at the high organic acid concentrations and low DO conditions associated with septicity.

The presence of large amounts of exocellular slime in activated sludge can be detected by two methods. A microscopic examination of the activated sludge "stained" with India ink will reveal large areas of the floc impenetrable to the India ink particles (Figure 4) in which cells can be seen surrounded by exocellular material. These exocellular slimes usually contain polysaccharides so that their presence and amount can be verified by the anthrone test for total sludge carbohydrates (Table 13). Activated sludge from domestic wastewater treatment plants normally contains 14% to 18% total carbohydrates (expressed as glucose on a dry weight basis by the anthrone test). Industrial wastewater treatment activated sludges may contain up to 20% to 22% carbohydrate expressed as glucose on a dry weight basis. In severe cases of "slimy sludge" or "viscous bulking," carbohydrate levels of up to 70% as glucose on a sludge dry weight basis have been seen. Interference in sludge settling starts to occur at an activated sludge carbohydrate content of approx-

imately 25% to 30% as glucose. Viscous bulking usually occurs when wastewaters high in readily metabolizable, soluble organics are treated under nutrient (N and/or P) deficient conditions. Apparently the exocellular slimes are products of a "shunt metabolism" or unbalanced growth, formed by the activated sludge microorganisms (both floc formers and filaments) when they cannot produce nitrogen- and phosphorus-containing cell material due to the lack of nutrients. Overproduction of "slime" has also been observed in domestic wastewater activated sludge plants not subject to nutrient deficiency. Here, slime production has been associated with very high oxygen uptake rates (often $> \sim 100 \text{ mg } O_2/g$ VSS, hr). These high metabolism rates probably lead to unbalanced growth because the bacteria are unable to obtain nutrients as fast as the organic materials can be processed. This type of problem has often been observed in domestic wastewater activated sludge plants using a high MLSS concentration and in intensely aerated aerobic digestion processes for wastewater activated sludges. In severe cases of slime bulking, the activated sludge is slimy to the touch and will hang in viscous "stringers" from the end of a sampling pipette. Often, the aeration basin or final clarifier will become covered with a viscous, sticky, solids-rich foam that can be in excess of several feet thick. In one severe case the slime "stuck" to the secondary clarifier sludge removal mechanism and would not flow into the RAS withdrawal line!! The slime material in this foam usually stains Neisser positive.

Other microscopic signs of nutrient deficiency include: (a) large clumps of Neisser positive staining cells in the flocs in the absence of an anaerobic/aerobic sequence in the aeration basin; (b) large amounts of intracellular PHB granules in the floc bacteria (and in the filaments); (c) unusual Neisser staining reactions of some filamentous organisms (e.g., type 0041, type 0675 and *H. hydrossis* often stain slightly Neisser positive when nutrient deficient); (d) extensive gonidia and rosette formation by *Thiothrix* spp. and type 021N (this can also occur under septic conditions); and (e) the activated sludge settles and compacts poorly even though the overall filamentous organism abundance is low ("viscous bulking").

The finding of algae (Figure 35f) in activated sludge is rare. Usually these organisms do not grow in activated sludge because there is not sufficient light in the aeration basin. The observation of algae indicates that there is another treatment unit in the process or that a recycle stream exists. For example, algae can grow on the top layers of biofilters and be flushed through the biofilter into an activated sludge unit in a coupled biofilter-activated sludge system. Algae can grow to large amounts in lagoons. Thus, if there is a supernatant recycle stream from a sludge lagoon back to the activated sludge process, algae can enter the activated sludge system in this way. The washing down of algae growth on secondary clarifier weirs can incidentally introduce algae into activated sludge. If the activated sludge system is followed by secondary effluent filters, then algae from the secondary clarifiers, as well as those growing on the filters themselves, can get back to the activated sludge in filter backwash if this is recycled through the activated sludge system. The presence of large amounts of algae in activated sludge has been caused by water treatment plant discharge of filter backwash to the collection system of the wastewater treatment plant.

PROTOZOA

Microscopic observation of protozoa and other higher life forms in activated sludge is a common and widespread practice. Generally, the types of organisms present can be related to plant performance and effluent quality. Further, these organisms are useful for assessment of toxicity.

From a morphological point of view, activated sludge is a relatively "simple" microbial community consisting of free and flocculated bacteria (and at times filamentous bacteria), protozoa, rotifers, nematodes, and a few other invertebrates. Protozoa and other higher life forms are usually aerobic and bacteriovorous (they eat bacteria). A few anaerobic flagellates and a number of saprophytic flagellates occur in activated sludge. The saprophytic flagellates use soluble organic matter for growth. Carnivorous protozoa, both free ciliates and attached ciliates (suctorians), occur which feed on other protozoa. Chlorophyll-bearing flagellates are incidentally observed, and are derived from the aeration basin walls.

Protozoa and other higher life forms constitute approximately 5% of the activated sludge biomass and are represented by about 200 species (Curds, 1973; Curds, 1975). Total numbers range from 100 to >100,000/mL. They are generally dominated by protozoa, with 500 to several thousand/mL observed commonly. These organisms perform several important functions in acti-

vated sludge, the most important of which is their removal of nonflocculated bacteria from wastewater through their feeding activities, yielding a clarified effluent (Curds et al., 1968; Curds and Fey, 1969). Additionally, these organisms may contribute to biomass flocculation through production of fecal pellets and mucus (Curds, 1975) and may function to break up large floc masses and encourage a more active biomass through their motility (Javornicky and Prokesova, 1963).

Observation of these organisms is easily done by placing one drop (0.05 mL) of activated sludge on a microscope slide, adding a cover glass, and examining it at 100X using phase contrast illumination. All protozoa and other higher life forms present should be counted by scanning the entire cover glass area using the mechanical stage of the microscope. The results of 4–5 separate preparations should be averaged. The total number of organisms present per mL of activated sludge culture will be equal to the average count per cover glass area times 20. If a large number of organisms are present (at times observed for flagellates), count the number present per field of view (at 100X), average this for 10–20 fields of view, and multiply this number by the number of fields of view for the cover glass area (typically about 300 at 100X magnification for a 22 × 22mm cover glass). This procedure should be conducted frequently (several times a week, or even daily) since protozoan populations in activated sludge can change rapidly in some circumstances (e.g., a toxicity upset).

Taxonomic classification of these organisms is based primarily on motility. The six basic groups observed in activated sludge are flagellates, amoebae, free-swimming ciliates, attached (stalked) ciliates, rotifers, and a few other invertebrates. Identification to species is not necessary; however, recognition of the major groups of protozoa and higher life forms is useful in activated sludge operation. The most common protozoan and higher life forms observed in activated sludge based on our experience are shown in Figures 36, 37, and 38. These photos can be used as a simplified identification "key." More detailed taxonomic identification keys can be found as follows: protozoa—Curds, 1969; Curds, 1975; Jahn et al., 1980; Mudrack and Kunst, 1986; rotifers—Calaway, 1968; Doohan, 1975; Gerardi, 1987; nematodes - Calaway, 1963; Schiemer, 1975; Tarjan et al., 1977; Gerardi, 1987a; and annelids—De L. G. Solbe, 1975. The major groups of protozoa and higher life forms found in activated sludge are described below.

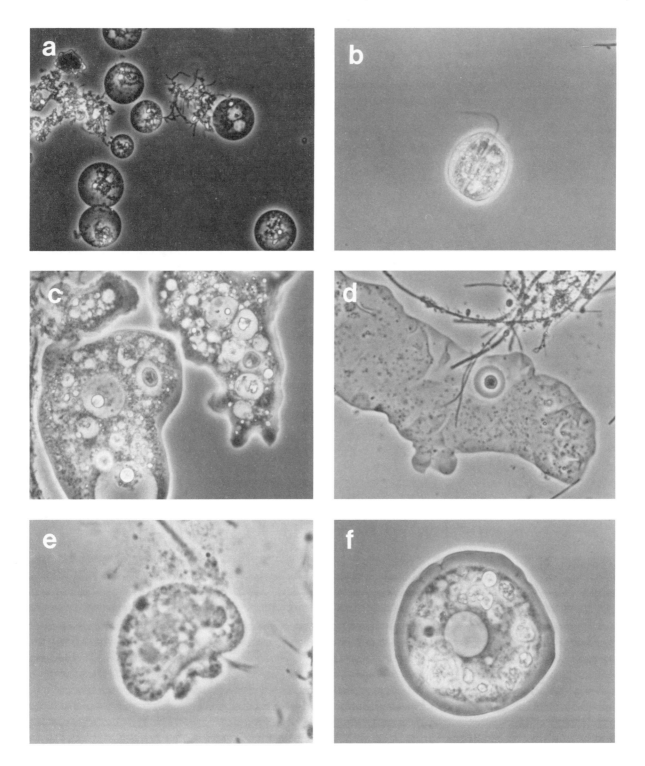

Figure 36. Common flagellates and amoebae found in activated sludge: *a. Monas* spp. (400X); *b. Trigonomonas* spp. (1000X); *c. Polychaos* spp. (400X); *d. Mayorella* spp. (400X); *e. Arcella* spp., *f.* top view (1000X) (all phase contrast).

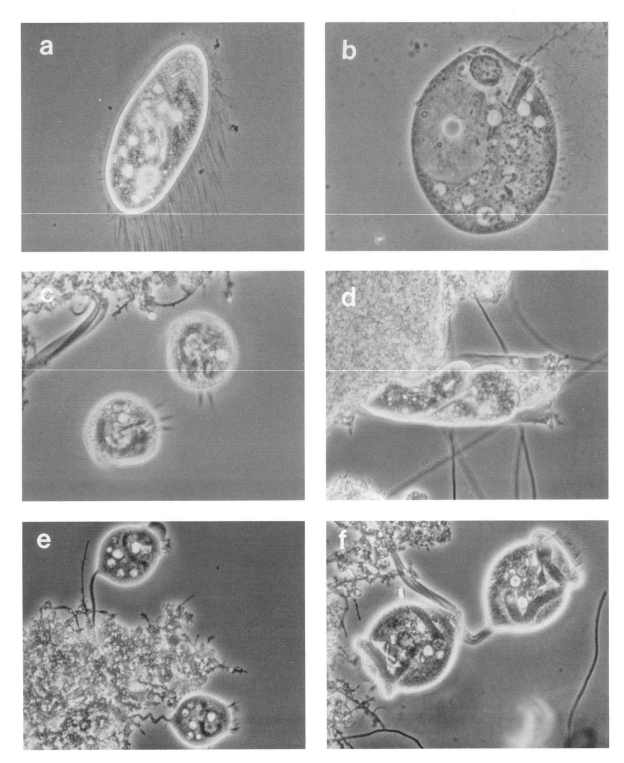

Figure 37. Common ciliates found in activated sludge: *a. Paramecium* spp. (200X); *b. Chilodonella* spp. (400X); *c. Aspidisca* spp. (400X); *d.Vaginicola* spp. (200X); *e. Vorticella microstoma* (400X); and *f. Vorticella campanula* (200X) (all phase contrast).

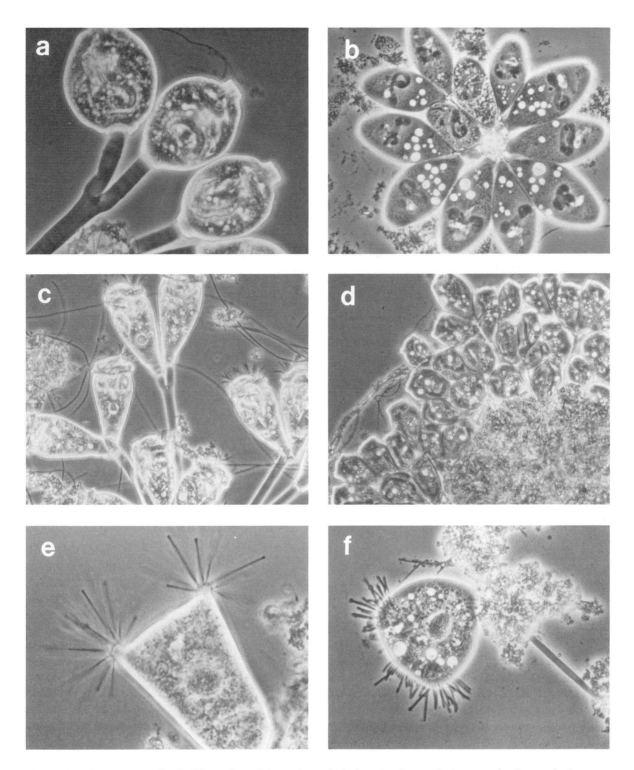

Figure 38. Common stalked ciliates found in activated sludge: *a. Opercularia* spp.; *b. Opercularia* spp.; *c. Epistylis* spp.; *d. Carchesium* spp.; *e. Tokophrya* spp.; and *f. Podophrya* spp. (all 200X phase contrast).

Flagellates: these are small (5–20 μm) oval or elongated forms, actively motile by one or more long, whip-like flagellae. Many species found in activated sludge feed on soluble organic matter, and their presence can indicate the presence of high soluble BOD levels.

Amoebae: these vary in shape and size (10–200 μm), and are motile by pseudopodia ("false feet"). Some species present in lightly loaded plants have a hard, ornate "shell." Amoebae grow well on particulate organic matter and are able to tolerate low DO environments.

Free-swimming ciliates: these are round- to oval-shaped (20–400 μm), and are actively motile by rows of short, hair-like cilia. Some species have cilia fused into spikes which aid them in crawling on the activated sludge suspended solids ("crawlers"). Ciliates usually occur under conditions of good floc formation and generally indicate good activated sludge operation.

Attached ciliates: these appear much as the ciliates above, but are found attached to flocs by a stalk which may be either rigid or contractile. Some species have one organism per stalk, while others are colonial. Stalked ciliates are generally a sign of stable activated sludge operation and the species found can be used to indicate approximate MCRT. The colonial forms occur at higher MCRTs.

Rotifers: these appear in various shapes and are much larger (50–500 μm) and have a more complex structure than protozoa (Figures 39a-e). Most are motile and attach to activated sludge flocs by a contractile "foot." These organisms occur over a wide range of MCRT; some species are indicative of high MCRT.

Higher invertebrates: these include nematodes (Figure 39f), tardigrades such as *Macrobiotus* (Figures 40c and d), and annelids such as *Nais* (Figure 40b) and *Aelosoma* (Figure 40a) (which can impart a reddish color to activated sludge due to their red- or orange-colored "eyespots"). Nematodes are generally observed only in higher MCRT systems, while the tardigrades and annelids appear to occur only in nitrifying activated sludge systems, probably due to their susceptibility to ammonia toxicity.

The various protozoan and invertebrate groups develop in activated sludge according to the growth conditions. Protozoa have maximum growth rates of one per day or higher at 20°C

(Curds, 1975), and rotifers have growth rates of 1–2 per day at 20°C (Doohan, 1975). Thus, growth rate rarely limits the development of these organisms in most activated sludge systems. Nematodes have a lower maximum growth rate and generally develop only in long MCRT systems. Food availability, principally freely dispersed bacteria or turbidity, is the primary determinant of which group predominates. Flagellates, amoebae, and some small, free-swimming ciliates require a high prey density ($> 10^6$–10^7/L) because their chase and capture-feeding mechanism is inefficient. These are selected for at plant startup and at low MCRT (high F/M) conditions. Attached ciliates, rotifers, and other invertebrates develop at lower prey densities because of their attachment to the activated sludge floc and their ability to feed by ciliary action ("filter feeding"). These organisms are selected for at high MCRT (low F/M). These factors lead to a marked succession of protozoa and other higher life forms in activated sludge according to process parameters (Table 20).

Optimum activated sludge performance occurs when there is a balance among free-swimming and attached ciliates and rotifers. An overabundance of flagellates, amoebae, or free-swimming ciliates is an indication of high F/M (low MCRT), while an overabundance of attached ciliates, rotifers, and other higher life forms, especially nematodes, is an indication of low F/M (high MCRT). These relationships are summarized in Table 20. Because sludge settling often deteriorates at the extremes of the F/M range, many plants attempt to adjust process parameters based on the types of protozoans and other higher life forms observed in the activated sludge to avoid these extremes of F/M. This is not a very sophisticated approach to activated sludge settleability control, because many other factors beside F/M contribute to the growth of the filamentous organisms that cause deterioration of activated sludge settling.

One of the most valuable uses of the microscopic observation of these organisms is for toxicity assessment. These organisms, particularly the ciliates and rotifers, are generally the first to be impacted by toxic materials and can serve as an *in situ* biomonitoring test for toxicants or other adverse stresses on the activated sludge process. The first noticeable sign of toxicity or stress is usually the slowing or cessation of cilia movement in the ciliates. Next, the predominant protozoan groups shift toward flagellates and small, free-swimming ciliates which often "bloom" to

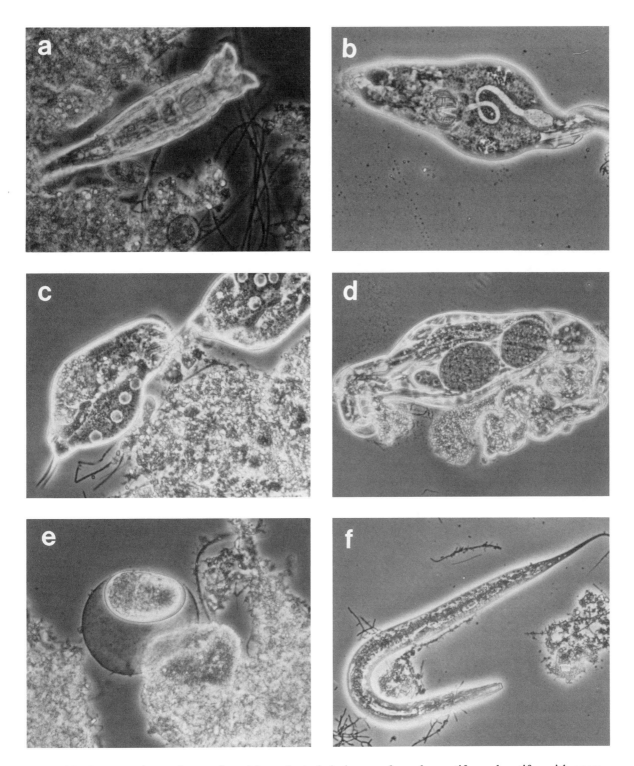

Figure 39. Common invertebrates found in activated sludge: *a.*, *b.* and *c.* rotifers; *d.* rotifer with eggs; *e.* rotifer cyst; *f.* nematode (all 200X phase contrast).

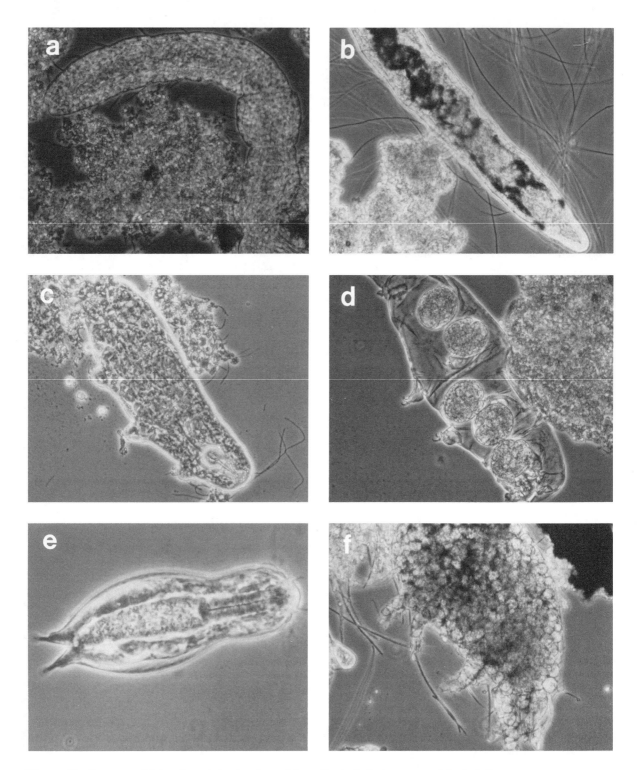

Figure 40. Common higher invertebrates found in activated sludge: *a*. and *b*. bristle worms; *a*. *Aeleosoma* spp. and *b*. *Nais* spp. (100X); *c*. and *d*. tardigrades (water bears) (*Macrobiotus* spp.; *d*. with eggs) (200X); *e*. gasterotrich (*Chaetonotus* spp. 200X); *f*. hydrachnid (water mite, 100X) (all phase contrast).

Table 20. Organic Loading of Activated Sludge and Predominant Higher Life Forms Observed[a]

Condition	Predominant Groups
High F/M; low MCRT	flagellates, amoebae, and small, free-swimming ciliates.
Moderate F/M; average MCRT	good diversity of organisms, dominated by free-swimming and stalked ciliates.
Low F/M; high MCRT	stalked ciliates, rotifers, and higher invertebrates, especially nematodes.

[a]From: Reynoldson, 1942; Baines et al., 1953; Curds and Cockburn, 1970 and 1970a; Curds, 1975; Mudrack and Kunst, 1986.

high numbers (>10,000 per mL). This is an indication of activated sludge floc breakup and the production of large numbers of dispersed bacteria (turbidity), which are utilized by the flagellates and free-swimming ciliates as a food source. Finally, in severe cases, these protozoans die, and their lysis and release of cell contents can cause activated sludge foaming (white foam containing dead protozoans and protozoan fragments). Toxicants that cause this chain of events include heavy metals and cyanide. Stresses other than toxicity that induce these responses include low DO, pH outside the range of 6.0 to 8.0, and high temperature. Protozoa and other higher life forms are generally absent from activated sludge systems operated at temperatures above 37–40°C.

FILAMENTOUS ORGANISMS

Microscopic examination of the types, abundance, condition, and growth forms of the filamentous organisms provides the greatest wealth of information about the nature and causes of solids separation problems in activated sludge. The ability to make conclusions of this kind from a microscopic examination of filamentous organisms is the result of many investigations, some of which have correlated observed filamentous organisms types in activated sludge with particular sets of wastewater characteristics, operating conditions and aeration basin configurations, and some of which have involved isolation and pure culture study of the filamentous organisms. We will discuss the results of the latter type of experiment in more detail in Chapter 3. Here we will concentrate on the application of the results of these investigations to problem diagnosis.

Before presenting specific information on problem diagnosis from filamentous organism

identification, a few general remarks are in order. It is rare that an activated sludge sample from a plant with bulking or foaming contains only one filamentous organism type. Usually three or more types are seen. Often there are one or more filamentous organism types present in significant (dominant) quantities (an abundance of "common" or greater), and several others present in smaller amounts (secondary). The correlation of the cause of bulking or foaming is made with the dominant filamentous organism types. The finding of several (or many) dominant filament types at one time is often an indication of unstable or varying conditions in the aeration basin. For example, low DO and low F/M filaments are at times found together. The filamentous organisms present in smaller amounts should not be ignored. Sometimes they can provide additional information such as the presence of a particular wastewater characteristic or treatment process, or the "left-overs" from a previous condition. Some filamentous organisms are quite specific indicators of a particular set of conditions, while others are present under a variety of conditions. In the latter case it is important to pay close attention to the condition of the filamentous organisms, since this can give clues to which one of several possible factors may be causing its growth. The factors influencing filamentous organism growth in activated sludge are just the same as those influencing the growth of all other types of microorganisms in activated sludge, and can be classified into "general factors" and "specific factors." The general factors are:

- MCRT or F/M
- Aeration Basin Configuration (or wastewater feeding regime)
- Presence of initial unaerated zones (anoxic or anaerobic) in the aeration basin where

the RAS and the influent wastewater mix together.

Richard (1989) presented information on the relationship of filamentous organism occurrence in activated sludge as a function of MCRT (F/M) for plants treating domestic wastes in Colorado, USA (Figure 41). It shows that some filamentous organisms occur over a fairly wide range of MCRT values (e.g., *S. natans*, type 1701, *Thiothrix* spp., and *Nocardia* spp.), while others are limited to a smaller range of MCRT. Types 0092, 1851, 0675, 0041, and *M. parvicella* are all associated with quite high MCRTs (≥ 10 days) .

The growth of many, but not all, filamentous organisms in activated sludge (e.g., type 021N, *Thiothrix* spp., *S. natans*, *N. limicola*, type 1701, *H. hydrossis*, type 1851, *Nocardia* spp.) is generally encouraged by the use of uniformly aerated, completely mixed, continuously fed aeration basins. When the aeration basin is compartmentalized to include an initial high F/M feed zone (a "selector"), and especially when this feed zone is anoxic or anaerobic, the growth of these filamentous organisms is suppressed. The same effect is seen when a true fed-batch mode of operation [the sequencing batch reactor (SBR)] is utilized with an unaerated feed and initial react period. The effect is not observed in systems that are operated in a mode where wastewater is fed over an extended period of time, then this is fol-

lowed by aeration, settling, and decanting. These types of systems are prone to the development of filamentous activated sludge just like continuous flow completely mixed systems.

There is a group of filamentous organisms that grow at high MCRT (for example, especially in nitrifying BNR plants) in the presence of initial unaerated (anoxic or anaerobic) zones. These are *M. parvicella*, type 0092, type 0041, and type 0675.

Specific factors influencing the occurrence of filamentous organisms in activated sludge are:

- DO concentration
- Nutrient (N and P) concentration
- pH
- Sulfide concentration
- Nature of organic substrate including whether it is soluble or particulate and whether it is readily biodegradable or slowly biodegradable
- Seeding from surfaces
- Surface trapping of foam and foam recycle

The following organisms are associated with low DO:

- type 1701
- *S. natans* } at low to moderate MCRT
- *H. hydrossis*
- *M. parvicella* } at high MCRT

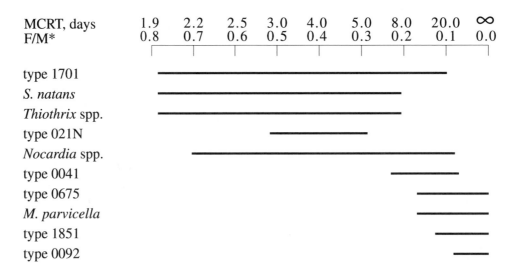

| MCRT, days | 1.9 | 2.2 | 2.5 | 3.0 | 4.0 | 5.0 | 8.0 | 20.0 | ∞ |
| F/M* | 0.8 | 0.7 | 0.6 | 0.5 | 0.4 | 0.3 | 0.2 | 0.1 | 0.0 |

*F/M as kg BOD5/kg MLVSS, day

Figure 41. Relationship of MCRT (and F/M) to the occurrence of specific filamentous organisms in activated sludge.

It will be seen in Chapter 3 that the phrase "low DO" is a relative term, since the DO concentration associated with low DO filamentous organism growth is a function of the F/M of the activated sludge.

N and P deficiency encourages the growth of type 021N, *Thiothrix* spp., type 0041, and type 0675. When growing under nutrient deficient conditions, type 021N and *Thiothrix* spp. often exhibit rosettes and gonidia (Figure 25e and Figures 26b, c and e), and they usually contain intracellular granules (most often PHB, but not sulfur). Types 0041 and 0675 growth under nutrient deficiency occurs almost uniquely in activated sludge treating industrial wastewaters. In nutrient-deficient activated sludge, types 0041 and 0675, which normally stain Neisser negative, often exhibit a Neisser positive exocellular slime covering. In addition to the observation of filamentous organisms, the detection of significant amounts of exocellular material in the activated sludge floc, using the India ink reverse-staining procedure and anthrone analysis, also may be used to diagnose a nutrient limitation.

The presence of large numbers of fungi in an activated sludge indicates low pH (pH < 6.0) and suggests that the influent contains strong acid. Some nitrifying activated sludge systems treating low alkalinity wastewater can exhibit low pH, as can oxygen activated sludge systems treating low alkalinity wastewater. Fungal bulking has never been observed under these circumstances. Fungi can be observed incidentally in the mixed liquor of coupled fixed-film/activated sludge systems, especially if there is not an intermediate clarifier between the two systems. Fungi grow well in fixed film reactors (especially biofilters), and they appear in the mixed liquor because they have washed out of the biofilter into the activated sludge system.

Several filamentous organisms can utilize sulfide as a source of energy (oxidizing it to sulfur and then depositing the sulfur as intracellular granules). These are *Thiothrix* spp., type 021N, *Beggiatoa* spp., and type 0914. When growing on sulfide, these organisms exhibit intracellular sulfur granules either in situ or in the S test (Table 9). Some of these organisms (*Thiothrix* spp. and type 021N; Richard et al., 1983; 1985) use both sulfide and the low molecular weight organic acids (e.g., acetic acid) produced by the fermentation processes that produce septic sewage and, thus, their growth is strongly encouraged by septicity.

Type 0914 is a fairly rare filamentous organism, with the exception that it seems to occur commonly in high MCRT, BNR-activated sludges.

Beggiatoa spp. occurs in fixed-film reactors and will enter activated sludge by washing through from the first stage of a coupled fixed-film/activated sludge process. *Beggiatoa* spp. is found very commonly in the heavy white tuft-like growth in the first stages of oxygen deficient (organically overloaded) RBC plants. The white color of this growth is due to the intracellular sulfur granules in the *Beggiatoa* spp. If one scrapes below the surface of the white growth, a black septic layer will be encountered. Thus the *Beggiatoa* spp. grows on the aerobic surface of a septic biofilm, obtaining sulfide from below and DO from the solution to oxidize the sulfide to sulfur.

Most of the filamentous organisms on which pure culture experiments have been conducted are able to grow well on fairly simple, soluble, readily metabolizable organic substrates. While this does not necessarily mean that these filamentous organisms grow on these substrates in the activated sludge culture, this information, taken together with in situ observation of filamentous organism occurrence, can be used to associate some filamentous organisms with the presence of different general types of substrate. Thus, the filamentous organisms that appear to be favored by soluble, readily metabolizable substrates are *S. natans*, type 021N, *Thiothrix* spp., *H. hydrossis*, *N. limicola,* and type 1851. The last two organisms are often seen when simple sugars are present in the wastewater.

The following filamentous organisms are apparently able to grow on slowly metabolizable (or perhaps even particulate) substrates: types 0041, 0675, and 0092, and *M. parvicella*. These four organisms were found to be the most common in BNR-activated sludge plants in South Africa (Blackbeard et al., 1988) which have initial anaerobic and/or anoxic zones. It is postulated (Ekama and Marais, 1986) that slowly metabolizable and particulate organic matter degradation rates are very slow in anaerobic and anoxic zones compared to the rates under aerobic conditions. Because of this, the particulate and slowly metabolizable substrates are transported to the aerobic zones of the BNR plants where their hydrolysis takes place, producing low concentrations of soluble organics which favor the growth of these filamentous organisms.

A study of the fate of filamentous organisms in aerobic digesters was made for 10 domestic wastewater activated sludge plants in Colorado (Richard, 1986). While most filamentous organisms present in the mixed liquor rapidly degraded and disappeared in the aerobic digester, the following five organisms were observed to grow: types 0041, 0675, 0092, *M. parvicella,* and *Nocardia* spp. *M. parvicella* and *Nocardia* spp. grew to problem levels in some aerobic digesters, causing foaming. The hydrolysis of particulate materials to produce low levels of soluble organics may be responsible for the growth of these filaments in aerobic digesters.

Type 0675 is almost always found in activated sludge treating pulp and paper wastewater. Richard and Jenkins (1985) found that type 0675 was the most common filamentous organism in this type of activated sludge.

We have previously made allusions to the occurrence of filamentous organisms in activated sludge as a result of their washing in from an upstream process (e.g., fungi, *Beggiatoa* spp., and *Thiothrix* spp.). This may be a fixed-film reactor in a coupled system or it may be a pretreatment unit discharging into the influent wastewater. In either case, the filamentous organisms entering the activated sludge will only grow in abundance if the appropriate growth conditions for them exist in the activated sludge system itself. Otherwise, they will be present only incidentally. Seeding of activated sludge units with filamentous organisms can be a serious problem in laboratory and pilot plant units. In these systems, the surface area to volume ratio of the unit itself, the feed lines, and the storage vessels is much higher than for prototype-activated sludge systems. Unless special care is taken, filamentous organisms such as *S. natans* and type 1701 will grow on the surfaces of the feed storage vessel and the feed lines, and will seed the activated sludge and cause bulking. This was first noticed by Gabb et al. (1989) when *S. natans* caused bulking in laboratory-scale nitrifying and BNR-activated sludge reactors, whereas this organism was never observed in full-scale systems operated under the same conditions and on the same wastewaters. A survey of the literature on pilot-scale activated sludge systems utilized for bulking control studies revealed the widespread occurrence of these "low DO" filamentous organisms under a wide range of operating conditions (Table 21). Seeding from growth in the feed system is suspected. The occurrence of this "artifact" can lead to erroneous conclusions on the effec-

tiveness of bulking control measures. Thus, if the activated sludge settles poorly because of *S. natans* growth (an experimental artifact), and measures are taken that improve the sludge settling, one will be led to the conclusion that these measures will provide an effective bulking control in a prototype system operated under the same conditions as the laboratory unit. This may not be the case, since all that the laboratory experiment achieved was to control an artifact of the experimental system!

These seeding problems can be prevented by regular cleaning of feed storage vessels, including scrubbing the walls with a sodium hypochlorite solution, and by regular cleaning of feed lines by immersion in, or flushing with, sodium hypochlorite solution. It is good practice to have a duplicate set of feed lines, with one set in service and the other set cleaned and ready to be placed in service. The activated sludge in these systems should be microscopically observed daily for the types of filamentous organisms that grow on surfaces.

Aeration basin and secondary clarifier design and scum or foam disposal techniques can strongly influence the levels of *Nocardia* spp. and *M. parvicella* in activated sludge. These organisms both have a tendency to float. Thus, any physical detail of the reactor system that tends to trap floating material, and any practice that recycles floating material back to the influent or the aeration basin will tend to enrich these organisms in the activated sludge. These factors are discussed in more detail in Chapter 4.

Using the information presented above it is possible to group most of the commonly observed filamentous organisms causing bulking into four broad groups. These groups were derived from ones proposed by Wanner and Grau (1989) and relate to both filamentous organism growth conditions in activated sludge as well as to potential control methods. These groups are presented in Table 22.

The utility of filamentous organism characterization in the diagnosis and subsequent rectification of bulking problems is illustrated in Table 23. Examples are given of wastewater treatment plants in which filamentous bulking occurred, in which the major causative filamentous organism(s) was identified, and from this the cause of bulking deduced. Based on this information, specific operation and/or design changes were made that successfully reduced the filamentous organism level and ameliorated the bulking problem.

Table 21. Conditions where *Sphaerotilus natans* was a Dominant Filamentous Organism in Laboratory-Scale Activated Sludge Units (Gabb et al., 1989)

Author	BOD/N	BOD/P	COD/N	COD/P	DO, mgO₂/L	MCRT, day	Aerator HRT, hr	F/M, gCOD/gVSS, day	F/M, gBOD/gVSS, day	Aerator MLSS, g/L
Clesceri (1963)	–	–	–	–	–	–	8	0.6[e]	–	–
Hattingh (1963)	≥22	≥168	–	–	–	–	23	–	–	–
Chudoba et al. (1973a)	–	–	–	–	–	–	8	1.1	–	–
Houtmeyers et al. (1980)	–	–	–	–	6	7	1	0.4[e]	–	4
Palm et al. (1980)[d]	–	–	–	–	0.1–5.5	1.9–11	–	0.38–1.72	–	–
Verachtert et al. (1980)	–	–	–	–	–	–	1	0.4	–	–
Chiesa and Irvine (1985)	–	–	–	–	–	–	8	0.56[e]	–	–
Lee et al. (1982)	6.8	–	16.3	–	4.5–16[e]	15	19	0.31–1.00	0.13–0.42	3–3.5
Rensink et al. (1982)	10	58	15.9	92.7	–	–	–	–	0.025–2.00[f]	2
van den Eynde et al. (1982a)	–	–	83	–	–	–	1	–	–	–
Wu et al. (1983)	20 & 69	–	95 & 28	–	7–8	–	24	0.2–0.4	0.1–0.3	1–3
Gabb et al. (1985)	–	–	–	–	5–8	12–14	2.5	–	0.1	2–3
van Niekerk (1985)	–	–	15.7 & 22.9	61 & 89	3.5–4.5	20–30	–	0.2–0.3	–	–
Burke et al. (1986)[c]	–	–	12.6	24	2–4	–	3.8	1.26	–	–
Still et al. (1986)[a]	–	–	5.4	–	1–7	20	24	0.35	–	1–2
Still et al. (1986)[b]	–	–	5.4	–	2–3	20	16	0.31	–	–

[a]Operated under completely aerobic conditions.
[b]Operated on a 4-hr cycle; e.g., 1 hr anoxic, 3 hr aerobic.
[c]Operated in high-rate Phoredox configuration. When one-third of the feed was bypassed to the aerobic reactor, S. natans grew in abundance; when all the feed was again fed to the anaerobic zone, S. natans abundance declined sharply.
[d]S. natans growth occurred at high DO only when a high F/M (low MCRT) prevailed, or at a low F/M (high MCRT) when the DO was at the lower end of the range given.
[e]F/M based on gCOD/gMLSS, day.
[f]F/M based on gBOD₅/gMLSS, day.

Table 22. Proposed Filamentous Organism Groups

Group I—Low DO Aerobic Zone Growers

Features
- readily metabolizable substrates
- low DO
- wide MCRT range

Organisms   *S. natans*, type 1701, *H. hydrossis*

Control
- aerobic, anoxic, or anaerobic selectors
- increase MCRT
- increase aeration basin DO concentration

Group II—Mixotrophic Aerobic Zone Growers

Features
- readily metabolizable substrates, especially low molecular weight organic acids
- moderate to high MCRT
- sulfide oxidized to stored sulfur granules
- rapid nutrient uptake rates under nutrient deficiency

Organisms   type 021N, *Thiothrix* spp.

Control
- aerobic, anoxic, or anaerobic selectors
- nutrient addition
- eliminate sulfide and/or high organic acid concentrations (eliminate septicity)

Group III—Other Aerobic Zone Growers

Features
- readily metabolizable substrates
- moderate to high MCRT

Organisms   type 1851, *N. limicola* spp.

Control
- aerobic, anoxic or anaerobic selectors
- reduce MCRT

Group IV—Aerobic, Anoxic, Anaerobic Zone Growers

Features
- grow in aerobic, anoxic (denitrification) and anaerobic (EBPR) systems
- high MCRT
- possible growth on hydrolysis products of particulates

Organisms   type 0041, type 0675, type 0092, *M. parvicella*

Control   largely unknown but:
- maintain uniformly adequate DO in aerobic zone and stage the aerobic zone

Table 23. Examples of "Curing" Bulking or Foaming by a Change in Full-Scale Plant Operation Suggested by Filament Identification

Treatment Plant; Wastewater Treated	Filamentous Organism(s) Causing Bulking or Foaming	Change in Operation and/or Design	Results
Plant A Domestic wastewater and fruit processing wastewater	*Thiothrix* spp. type 1701	• Increase NH$_3$ supplementation • Increase aeration; use liquid O$_2$	*Thiothrix* spp. eliminated, type 1701 remained Type 1701 reduced; return to nonbulking condition
Plant B Pulp and papermill wastewater	*H. hydrossis* type 0803	• Increase aeration • Increase sludge wasting rate	*H. hydrossis* eliminated Type 0803 eliminated; return to nonbulking condition
Plant C Domestic wastewater			
(a) winter operation	type 1701	• Increase aeration	Type 1701 eliminated; return to nonbulking condition
b) summer operation (lower F/M for nitrification)	*M. parvicella*	• Increase sludge wasting rate	*M. parvicella* eliminated; return to nonbulking condition
Plant D Domestic and soluble industrial wastewater mixture	type 0675 type 0041	• Change from step feeding to "plug flow"	Types 0675 and 0041 reduced; return to nonbulking condition
Plant E			
Domestic + industrial wastewater; oxygen activated sludge (first-stage DO uptake rate = 200 mgO$_2$/gVSS, hr)	type 1701	• Increase DO concentration from 10–12 mg/L to 16–20 mg/L in first stage of aeration basin	Type 1701 eliminated; return to nonbulking condition
Plant F			
Domestic + pulp and paper and other industrial wastewater	type 1851 and *N. limicola* II, types 0675 and 0041	• Install aerobic selector	Filament abundance decreased; nonbulking condition achieved; dominant filamentous organisms changed to types 0675 and 0041

Table 23. Continued

Treatment Plant; Wastewater Treated	Filamentous Organism(s) Causing Bulking or Foaming	Change in Operation and/or Design	Results
Plant G			
Pulp and paper wastewater	type 021N	• Provide aeration for flow equalization basin to remove septicity	Type 021N eliminated; return to nonbulking condition
Plant H			
Domestic wastewater	M. parvicella	• Decrease MCRT • Increase aeration basin DO	M. parvicella eliminated; return to nonbulking condition
Plant I			
Domestic wastewater; single stage nitrification	M. parvicella	• Install aerobic selector and add diffused aeration basins during expansion	M. parvicella eliminated; return to nonbulking condition
Plant J			
Domestic plus poultry processiong wastewater	type 1701 and S. natans	• Convert from complete mix aeration basin to anaerobic selector activated sludge system	Type 1701 and S. natans eliminated; return to nonbulking condition
Plant K			
Domestic wastewater	S. natans, Thiothrix spp.	• Convert from complete mix aeration basin to anoxic selector system	S. natans and Thiothrix spp. eliminated; return to nonbulking condition
Plant L			
Domestic wastewater	type 1701	• Combined RAS and a high DO mixed liquor recycle stream prior to recycle to deoxygenate prior to anoxic zone	Type 1701 eliminated; return to nonbulking condition

CHAPTER 3

Control of
Activated Sludge Bulking

GENERAL APPROACH FOR CONTROL OF FILAMENTOUS ORGANISMS IN BULKING ACTIVATED SLUDGE

It almost goes without saying that the best approach for preventing the growth of excessive levels of filamentous organisms in activated sludge and ensuring long-term operation free from sludge settling problems is through appropriate design and operating methods. However, since many of these techniques have been developed only recently, many existing activated sludge plants are faced with solids separation problems. For the correction of these existing problems the following general approach is recommended:

1. Use the microscopic examination procedure outlined in Chapter 2 to identify the causative filamentous organism(s).
2. Use this identification and the information given in this manual, together with a knowledge of the plant operating conditions and wastewater characteristics, to determine the probable cause(s) of the filamentous organism(s) growth.
3. If the probable cause is one that can be rectified without major changes, make the required operational changes to address the problem. For example, if septic wastewater is indicated, wastewater prechlorination may be initiated. If the probable cause is nutrient deficiency, determine which nutrient(s) is deficient by analysis of influent and effluent and rectify the deficiency by increasing the feed rate of existing nutrient supply system(s) or by installing nutrient addition facilities.
4. Some plants with bulking sludge problems may require major design or operational changes that can take a long time to implement (e.g., additional aeration capacity,

changes in aeration basin configuration, industrial waste control, etc.). In addition, once changes have been made to discourage filamentous organism growth, sludge settleability may improve only slowly. It should be recognized that, following a change in operational conditions to disfavor filamentous organism growth, the activated sludge microbial population changes at a rate proportional to the culture washout rate (or mean cell residence time, MCRT). Thus, in a completely mixed system, a period equal to approximately 3 MCRT is required for a bulking activated sludge to return to a nonfilamentous condition (Palm et al.,1980). For a plant with an MCRT of 10 days, approximately one month would be required to return to a nonbulking condition. Some data on the amelioration of low F/M bulking by the installation of "selectors" suggests that, in some cases, even longer than 3 MCRTs may be required for this type of plant modification to exert its full effect on sludge settleability (Wheeler et al., 1984). Because of these factors, rapid, nonspecific methods may be needed to eliminate the symptoms of bulking. These methods fall into three categories:

a. Manipulation of RAS flow rates and wastewater feed points to the aeration basin.
b. Addition of chemicals to enhance the settling rate of the activated sludge without attempting to selectively limit the growth of filamentous organisms.
c. Addition of toxicants to the activated sludge to selectively kill the extended filamentous organisms that cause bulking.

The last two methods cited above do not change the conditions that caused the bulking, so that the bulking will most likely return when these control methods cease.

MANIPULATION OF RAS FLOW RATES AND AERATION BASIN FEED POINTS

The adverse effects of a bulking sludge often can be minimized by proper management of the activated sludge system, particularly if it is underloaded and/or certain process options were built into the original design. To understand how these process management tools can be used to deal with bulking sludge it is necessary to:

- Review activated sludge secondary clarifier operating principles, particularly as they relate to the clarifier thickening function
- Present techniques for determining the SS handling capacity of a secondary clarifier, and how this is influenced by sludge settling characteristics
- Review techniques of secondary clarifier and aeration basin operation for maximizing secondary clarifier capacity in activated sludge systems operating with bulking sludge.

CLARIFIER OPERATION PRINCIPLES

Activated sludge secondary clarifiers have two basic functions—clarification and thickening. Clarification is the removal of activated sludge flocs to produce a clear overflow that meets discharge standards and does not overload downstream processes (e.g., tertiary filtration). Thickening of the settled activated sludge SS (i.e., RAS) is required so that they can be returned to the aeration basin. If the activated sludge SS are not thickened and removed from the secondary clarifier at a rate faster than they are added they will accumulate until the secondary clarifier is full, and the excess SS will be discharged into the effluent. Since bulking affects the ability of the activated sludge SS to thicken, it is the thickening function and thickening capacity of the secondary clarifier that is affected by sludge bulking.

PROCESS SCHEMATIC AND DEFINITIONS

Figure 42 is a process schematic of the activated sludge system. The aeration basin has a volume V, and the secondary clarifier has a surface area A. Wastewater enters the aeration basin at flowrate Q, together with RAS, at flowrate Q_r. The aeration basin SS concentration (MLSS) is X, and the RAS, SS is X_u. Activated sludge SS are applied to the secondary clarifier at a flowrate of $Q + Q_r$.

It is assumed here that sludge is wasted from the RAS at a flowrate Q_w, at the same SS concentration as the RAS, X_u. Effluent is discharged from the secondary clarifier at a flowrate of $Q-Q_w$, and it contains an SS concentration of X_e.

PROCESS OPERATING RELATIONSHIPS

In developing relationships to describe the secondary clarifier thickening function, certain simplifying assumptions can be made. First the WAS flowrate, Q_w, usually is small compared to the influent flowrate, Q, and its effect on effluent flow can be neglected. Second, the quantities of activated sludge SS discharged in WAS and in the secondary effluent are small compared to the quantity of activated sludge SS applied to and withdrawn from the secondary clarifier.

Using these assumptions and assuming that SS are not accumulating in the secondary clarifier, a simplified SS mass balance over the secondary clarifier is:

$$(Q + Q_r)X = Q_r X_u \qquad (3.1)$$

Equation 3.1 states that, at steady-state, the rate at which SS are applied to the secondary clarifier must be equal to the rate at which they are removed. Equation 3.1 can be used to develop the following relationships that are useful for assessing secondary clarifier operation:

Degree of Thickening Achieved by a Secondary Clarifier

$$\frac{X_u}{X} = \frac{(Q + Q_r)}{Q_r} \qquad (3.2)$$

This relationship illustrates that the degree of thickening is a function of the influent and RAS flowrates. With the limits described below, the degree of thickening is determined by the plant operator when a particular value of Q_r is selected.

Calculation of Required RAS Flowrate

$$Q_r = \frac{QX}{(X_u - X)} \qquad (3.3)$$

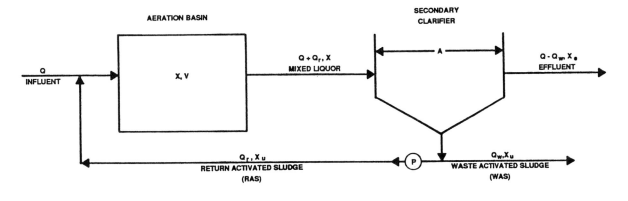

Figure 42. Activated sludge process schematic.

This relationship can be used to calculate the RAS flowrate required for a specified influent flowrate, the current aeration basin MLSS concentration, and an RAS, SS concentration that it is believed (perhaps based on past history) can be achieved in the secondary clarifier.

Secondary Clarifier Capacity Calculation

$$Q = \frac{Q_r(X_u - X)}{X} \qquad (3.4)$$

This relationship can be used to calculate the capacity of a secondary clarifier for specified values of RAS flowrate, aeration basin MLSS concentration, and achievable RAS, SS concentration. Equations 3.1 to 3.4 assume that sufficient secondary clarifier surface area is available to thicken the RAS to a concentration of X_u. Techniques for determining the secondary clarifier surface area required to achieve this are discussed in the following section.

SLUDGE THICKENING THEORY

Thickening of activated sludge SS in a secondary clarifier occurs by gravity settling. As the SS settle, their concentration increases from the MLSS concentration (X) to the RAS concentration (X_u). Because of their flocculent nature and concentration, activated sludge particles do not settle individually; rather, they settle as an entire mass. The

SS settle at a constant rate until they begin to "pile up" at the bottom of the container in which they are settling. The initial settling velocity of the activated sludge SS (V_i) decreases as the initial SS concentration increases. This type of behavior is called zone settling, or Type III settling.

Zone settling is familiar to anyone who has run an SVI test. The entire mass of activated sludge SS settles together, producing a well-defined interface between the top of the settling sludge and the clear supernatant. When the height of the interface is plotted against time, the line initially is straight but later begins to level off (Figure 43). The slope of the initial straight line is the initial settling velocity, V_i. Activated sludges with higher initial SS concentrations settle slower that those with lower initial SS concentrations (Figure 43).

Secondary clarifier thickening capacity is related to the ability to accept the applied SS load and to convey all the SS to the underflow. This is accomplished in two ways: settling of the activated sludge SS, and withdrawal of RAS by pumping. RAS withdrawal causes a general flow of fluid to the bottom of the secondary clarifier, and this general flow also carries the settling activated sludge SS to the bottom, where they can be removed.

The various techniques that have been developed to determine the thickening capacity of secondary clarifiers generally compare the rate at which activated sludge SS are applied to the rate at which they are conveyed to the bottom of the

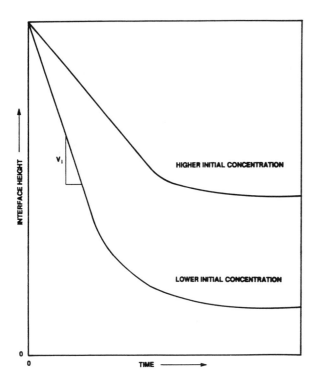

Figure 43. Zone settling of activated sludge suspended solids.

secondary clarifier by settling and bulk withdrawal. As long as the application rate is no greater than the settling rate plus the withdrawal rate, the applied SS will be conveyed to the bottom of the secondary clarifier and removed (i.e., thickening failure will not occur). If the SS application rate is greater, then excess SS will accumulate in the secondary clarifier until it is full and SS will overflow into the secondary effluent.

Techniques for secondary clarifier thickening analysis are relatively complex. They account for SS concentration increases in the secondary clarifier during settling and the effects of this on the rate at which SS are conveyed to the bottom of the secondary clarifier by both settling and bulk withdrawal. These techniques are discussed thoroughly in the literature (Keinath, 1977); they will not be discussed in detail here.

The thickening capacity of a secondary clarifier can be stated in terms of an allowable applied SS loading rate (G), which is the mass of SS applied per unit time per unit of secondary clarifier surface area. Using the definitions presented earlier, the actual SS loading rate is:

$$G_a = \frac{X(Q + Q_r)}{A} \qquad (3.5)$$

Comparison of the actual SS loading rate with the allowable rate will indicate whether the secondary clarifier is overloaded and whether it will experience thickening failure.

The allowable SS loading rate is a function of several factors including the RAS flowrate per unit of clarifier surface area (Q_r/A) and the settling characteristics of the activated sludge SS. By varying these factors, it may be possible to increase the capacity of the secondary clarifier and/or of the activated sludge system.

Secondary clarifier thickening capacity can be determined be direct measurement or by calculation using the techniques discussed below.

Direct measurement of thickening capacity is accomplished by manipulating the secondary clarifier SS loading until thickening failure occurs. For example, the secondary clarifier influent flowrate can be increased while the RAS flowrate is maintained constant. The secondary clarifier sludge blanket depth is monitored, and the point at which it begins to increase corresponds to the point at which thickening failure is just beginning to occur. The applied SS loading rate under these conditions is calculated. This is the allowable SS loading rate for the selected RAS rate and the observed sludge settling characteristics. By repeating this procedure at different RAS rates and with activated sludges of different settling characteristics, the secondary clarifier thickening capacity can be established for a variety of conditions.

While this technique has been used in practice, it is rather cumbersome and time-consuming and the secondary effluent quality may be degraded during testing. Moreover, bulking sludge must be available to allow the secondary clarifier capacity to be measured for bulking conditions!! The major advantage of this method is that the determination of allowable thickening capacity is direct and no assumptions are required to translate the results to full-scale plant performance. It may best be applied retroactively to the analysis of a sludge bulking incident that resulted in secondary clarifier failure. Documentation of the secondary clarifier SS loadings which resulted in thickening failure gives values which can be used to develop operating strategies to avoid secondary clarifier failure in subsequent bulking incidents.

Calculation techniques for determining allowable secondary clarifier thickening capacity require the measurement of sludge settling characteristics. The relationship between SS concentration and the initial settling velocity must be

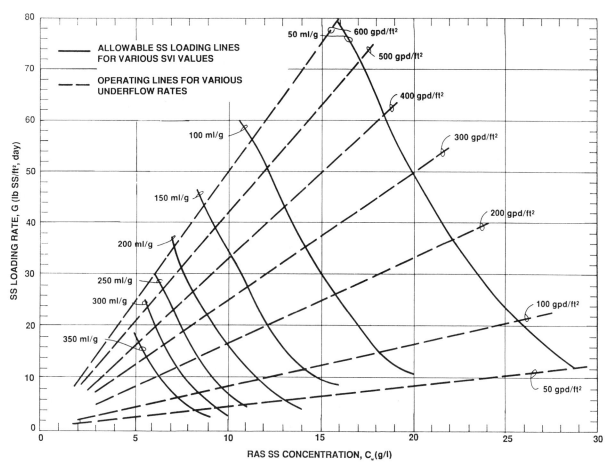

Figure 44. Secondary clarifier design and operation diagram. (Daigger and Roper, 1985). Reprinted with permission of the Water Environment Federation.

established, either by direct measurement or by using correlations between an index of sludge settleability and the SS concentration/initial settling velocity relationship developed by others. Correlations have been developed using indices such as the standard SVI, the stirred SVI conducted at MLSS = 3.5 g/L ($SSVI_{3.5}$) and the DSVI.

Direct measurement of the SS concentration/ initial settling velocity relationship can be quite cumbersome and time-consuming. Initial settling velocities must be measured for activated sludges with a range of initial SS concentrations obtained by mixing various proportions of mixed liquor, RAS, and secondary effluent. Measurements must be made in settling columns that are at least 0.15 m (6 in.) in diameter, 1.8 m (6 ft) tall, and equipped with a stirring device. Bulking sludge must be available before its thickening characteristics can be measured.

Correlations such as those discussed above can

be coupled with standard thickening capacity calculations to produce a secondary clarifier operating diagram. Figure 44 presents the results of one such correlation developed by Daigger and Roper (1985) in which the allowable SS loading rate (G_a) is plotted as a function of the RAS, SS concentration for sludges with SVIs in the range 50–350 mL/g. Dashed lines that correspond to various underflow rates (Q_r/A) also are plotted. To use this diagram, the point that corresponds to the clarifier operating conditions is located using two of the following pieces of information:

1. the actual SS loading rate (G)
2. the underflow rate (Q_r/A)
3. the RAS, SS concentration (X_u)

If all of these data are available, then one piece of information can be used as a check. When the point plots below the allowable SS loading line of interest, the secondary clarifier is operating below its allowable thickening capacity and should not be subject to thickening failure. When the point falls directly on the allowable SS loading line of interest, the secondary clarifier is operating at the failure point. When the point plots above the allowable SS loading, thickening failure will occur. An example of the use of this diagram is presented later in this section.

Correlations such as the one presented in Figure 44 have been developed by several workers (White, 1976; Johnstone, et al., 1979; Pitman, et al., 1984; Koopman and Cadee, 1983; Daigger and Roper, 1985; and Wahlberg and Keinath, 1988). All of these correlations have been successfully applied to the analysis of secondary clarifiers at full-scale wastewater treatment plants. Two factors must be remembered in the application of any of these correlations. First, all correlations are empirical and, consequently, they will have a limited range of application. Whenever possible, the correlation used in a particular situation should be "calibrated" by comparing actual secondary clarifier thickening performance with the predictions of the empirical correlation. Second, the various indices of activated sludge settleability used in each of these correlations are not equivalent. Table 24 summarizes the characteristics of the four most commonly used indices. In Chapter 1 it was shown that the indices vary in their correlation with filament populations and sludge thickening characteristics. While the DSVI

seems to provide the most accurate estimate of activated sludge settling characteristics, current practice in the USA (unfortunately) relies heavily on the use of the conventional SVI technique.

In spite of these difficulties, correlations between activated sludge settleability and settling characteristics have been used successfully to analyze full-scale wastewater treatment plants. Although care should always be exercised, and the relationships should be calibrated whenever possible, they can be used with confidence to estimate secondary clarifier capacity.

SECONDARY CLARIFIER ANALYSIS AND OPERATION

The first steps in secondary clarifier analysis often are to use the relationships in Equations 3.1 to 3.4. For example, sludge bulking generally will cause a decrease in the RAS, SS concentration. Equation 3.1 predicts that a decrease in the RAS, SS concentration (X_u) requires an increase in RAS flowrate. This increase is necessary to ensure that sludge applied to the secondary clarifier is removed in the RAS.

For example, consider an activated sludge system that operates at an influent flow of 1.0 MGD (0.044 m³/sec) and an RAS flow of 0.33 MGD (0.014 m³/sec). The MLSS concentration is 3000 mg/L, and the RAS, SS concentration is 12,000

Table 24. Comparison of Sludge Settleability Indices

Index	Settling Device	Low-speed Stirring	SS Concentration
Conventional SVI (SVI)	1-L Graduate cylinder	No	Aeration basin mixed liquor
Mallory SVI (SVI_m)	Mallory settleometer	No	Aeration basin mixed liquor
Diluted SVI (DSVI)	1-L Graduate cylinder	No	Mixed liquor diluted so that settled volume in graduate cylinder is less than or equal to 200 mL/L
Stirred SVI at 3.5 gSS/L ($SSVI_{3.5}$)	1-L Graduate cylinder	Yes	Tests at several SS concentrations; value at 3.5 g/L obtained by interpolation.

mg/L. As a check, calculate the required RAS flowrate using Equation 3.1.

$$Q_r = QX/(X_u - X)$$
$$= (1.0 \text{ MGD})(3,000)/(12,000-3,000)$$
$$= 0.33 \text{ MGD } (0.014 \text{ m}^3/\text{sec})$$

This agrees with the actual RAS flowrate. Now assume that the sludge settleability deteriorates and that the secondary clarifier sludge blanket begins to increase. The RAS, SS concentration is found to be 8,000 mg/L. Equation 3.1 indicates that the RAS flowrate must be:

$$Q_r = QX/(X_u - X)$$
$$= (1.0 \text{ MGD})(3,000)/(8,000-3,000)$$
$$= 0.60 \text{ MGD } (0.026 \text{ m}^3/\text{sec})$$

Thus, for these conditions, increasing the RAS flowrate to 0.60 MGD (0.026 m^3/sec) will be necessary to prevent secondary clarifier failure.

The use of Equations 3.1 to 3.4 to develop alternative secondary clarifier operating strategies must be accompanied by an analysis of secondary clarifier thickening capacity. In this manual the correlation method (with SVI) of Daigger and Roper (1985) will be used to illustrate this analysis. As indicated previously, other somewhat less convenient techniques also are available.

Assume that the secondary clarifier in the example discussed above is 45 ft (13.7 m) in diameter and has a surface area of 1,590 ft^2 (148 m^2). From Equation 3.5, the applied SS loading rate for the initial operating condition is:

$$G_a = \frac{X(Q + Q_r)}{A}$$
$$= (3,000 \text{ mg/L})(1.0 + 0.33)\text{MGD}(8.34)$$
$$/1,590 \text{ ft}^2$$
$$= 20.9 \text{ lb/ft}^2,\text{day}(102 \text{ kg/m}^2,\text{day})$$

This condition plots as point number 1 in Figure 45 and indicates that the secondary clarifier could operate successfully with a sludge having an SVI of up to 150 mL/g.

For the second operating condition (RAS, SS = 8,000 mg/L, RAS flowrate = 0.6 MGD), the applied SS loading rate is:

$$(3,000 \text{ mg/L})(1.0 + 0.6) \text{ MGD } (8.34)/$$
$$1,590 \text{ ft}^2 = 25.2 \text{ lb/ft}^2,\text{day } (123 \text{ kg/m}^2,\text{day})$$

This condition plots as point number 2 in Figure 45 and indicates that the secondary clarifier now can operate successfully with a sludge having an SVI just over 200 mL/g.

Now assume that the maximum RAS pumping capacity is 0.95 MGD (0.042 m^3/sec). At the maximum RAS flowrate the bulk withdrawal rate is 0.95 MGD/1,590 ft^2 or 597 gpd/ft^2 (24.3 m/day) and the applied SS loading rate is:

$$(3,000 \text{ mg/L})(1.0 + 0.95) \text{ MGD } (8.34)/(1,590$$
$$\text{ft}^2 = 30.7 \text{ lb/ft}^2,\text{day } (150 \text{ kg/m}^2,\text{day})$$

This condition plots as point number 3 on Figure 45 and indicates that the secondary clarifier could be operated successfully with a sludge having an SVI of about 250 mL/g and that the RAS, SS concentration would be about 6,300 mg/L.

Now assume that the RAS flowrate is maintained at 0.95 MGD (0.042 m^3/sec), equivalent to a bulk withdrawal rate of 597 gpd/ft^2 (24.3 m/day), but that the SVI will be controlled to a value of no more than 150 mL/g by RAS chlorination (see later sections in this chapter). Moving along the operating line (for 597 gpd/ft^2) in Figure 45 to point number 4 indicates that the secondary clarifier could be operated at an SS loading rate of up to 42.8 lb/ft^2,day (209 kg/m^2,day) and with an RAS, SS concentration of up to 8,700 mg/L at this loading.

Equation 3.5 can be rearranged to calculate allowable influent flow and allowable MLSS for these conditions:

- the maximum allowable influent flow rate at MLSS = 3000 mg/L

$$Q = (G_a A/X) - Q_r$$
$$= (43.5 \text{ lb/ft}^2,\text{day})(1,590 \text{ ft}^2)/((3,000 \text{ mg/L})$$
$$(8.34)) - 0.95 \text{ MGD}$$
$$= 1.9 \text{ MGD } (0.079 \text{ m}^3/\text{sec})$$

- the maximum allowable MLSS at an influent flow of 1.0 MGD (0.942 m^3/sec)

$$X = (G_a A)/(Q + Q_r)$$
$$= (43.5 \text{ lb/ft}^2,\text{day})(1,590 \text{ ft}^2)/((1.0 + 0.95)$$
$$\text{MGD } (8.34))$$
$$= 4,250 \text{ mg/L}$$

These examples illustrate how the general principles of secondary clarifier analysis can be used to optimize the existing operations.

System Analysis and Operation

In addition to manipulation of RAS flowrates, the effects of bulking sludge can be ameliorated by reducing MLSS concentration in the secondary clarifier feed. This reduces the SS load to the

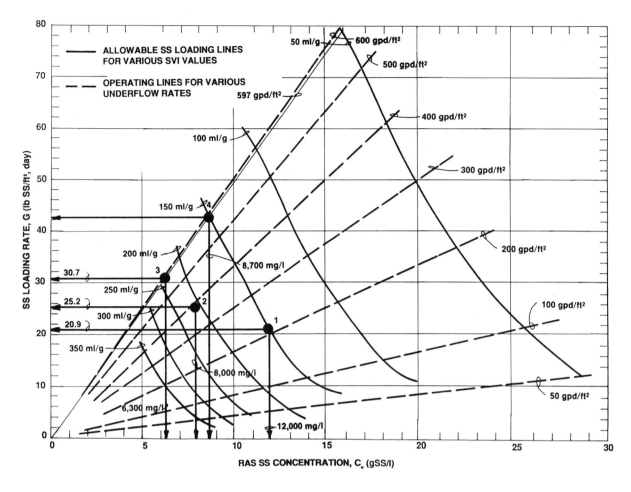

Figure 45. Secondary clarifier thickening analysis example.

secondary clarifier (Equation 3.5). Also reducing applied SS loading rate while maintaining the same RAS bulk withdrawal rate (i.e., move downward and to the left along an RAS bulk withdrawal operating line in Figure 45) increases the allowable SVI.

MLSS concentration in the secondary clarifier feed can be reduced by reducing the MLSS inventory or by changing the activated sludge operating mode. MLSS inventory can be reduced by increasing the activated sludge wasting rate. This may not be feasible if sludge handling capacity is not available, or if reducing the MLSS inventory would lead to an unfavorable F/M (e.g., a low F/M may be required to achieve nitrification).

The objective of changing operational mode to combat sludge bulking is to reduce MLSS concentration in the secondary clarifier feed without reducing the MLSS inventory. The step feed (step aeration) and contact stabilization configurations are particularly useful in this regard. Needless to

say, they can only be employed to combat bulking sludge if the plant is designed with the flexibility to operate in these configurations. Both configurations allow the operation of part of the aeration basin at a higher MLSS concentration (for a given F/M) than would be achieved were the aeration basin completely mixed.

The use of step feed to lower the SS loading applied to the secondary clarifier will be illustrated using the previous example. In addition to an influent flowrate (Q) of 1.0 MGD (0.042 m³/sec), an RAS flowrate (Q_r) of 0.33 MGD (0.014 m³/sec), an MLSS concentration (X) of 3,000 mg/L, an RAS, SS concentration (X_u) of 12,000 mg/L, and a secondary clarifier SS loading rate (G_a) of 20.9 lb/ft²,day (102 kg/m²,day), it is assumed that the aeration basin has a total volume of 0.25 MG (3780 m³) and that it can be operated in either the plug flow (conventional) or the two-pass step feed mode (Figure 46).

Point number 1 in Figure 47 indicates that

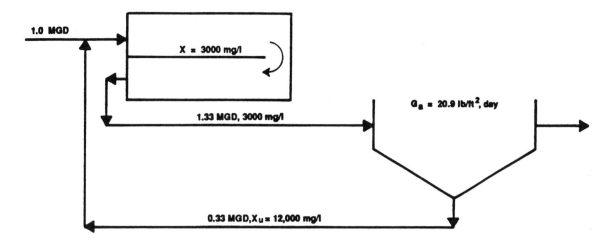

"PLUG FLOW" (CONVENTIONAL) MODE

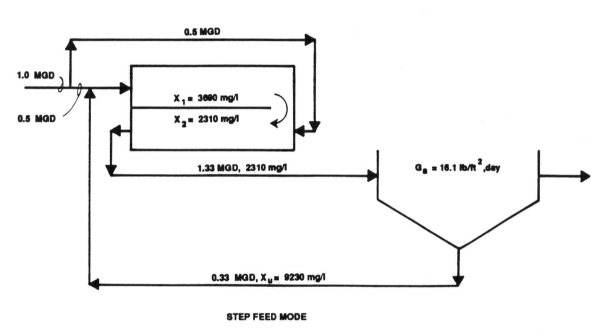

STEP FEED MODE

Figure 46. Schematics of "plug flow" and step feed operation examples.

these operating conditions are acceptable when the SVI is less than 150 mL/g.

When the operating mode is changed to two-pass step feed at the same influent and RAS flow rates, all of the RAS, but only one-half of the influent flow, is added to the first pass. The remainder of the influent flow is added to the second pass. A redistribution of mixed liquor SS inventory occurs so that it is higher in the first pass than in the second pass. This arises because less dilution of the RAS by influent occurs in the first pass than in the second pass. If the total

MLSS inventory is maintained the same as in the plug flow mode, then the MLSS concentration in the second pass will be less than when the system was operated in the plug flow mode.

The MLSS concentrations in each of the two passes can be calculated for step feed by writing SS mass balances for each pass. These equations then are substituted into an equation for the total SS inventory, which is solved for the SS concentrations in each pass and in the RAS. For our example the aeration basin MLSS inventory in the plug flow mode is:

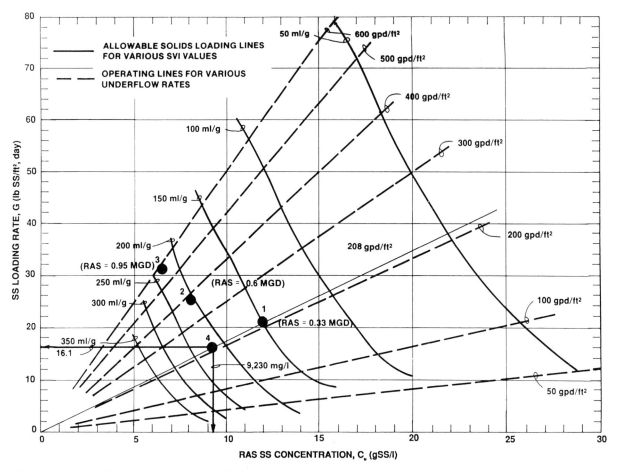

Figure 47. Step feed operation example (see text).

Aeration Basin SS Inventory = (3,000 mg/L)(0.25 MG)(8.34) = 6255 lb (2840 kg)

This SS inventory will be maintained in the two-pass step feed mode.

The SS mass balance for the point at which the influent flow to the first pass (0.5 MGD) is mixed with the RAS flow (0.33 MGD) is:

$$(0.33 \text{ MGD})(X_u) = (0.5 + 0.33)(\text{MGD})(X_1)$$

$$X_1 = (0.33/0.833)X_u = 0.4 \ X_u \quad (3.6)$$

The SS mass balance for the point at which the flow from the first pass (0.83 MGD) mixes with the influent flow to the second pass (0.5 MGD) is:

$$(0.83 \text{ MGD})(X_1) = (0.83 + 0.5)(\text{MGD})(X_2)$$

and

$$X_2 = (0.83/1.33) \ X_1 = 0.625 \ X_1 \quad (3.7)$$

Combining Equations 3.6 and 3.7 gives:

$$X_2 = 0.625 \ X_1 =$$
$$0.625 \ (0.4 \ X_u) = 0.25 \ X_u \quad (3.8)$$

The total aeration basin MLSS inventory can be expressed as a function of X_u as follows, remembering that the volume of each aeration basin pass is 0.25 MG/2 or 0.125 MG (1890 m³):

Aeration Basin Suspended Solids Inventory
$$= (X_1)(0.125 \text{ MG})(8.34) +$$
$$(X_2)(0.125 \text{ MG})(8.34)$$
$$= 1.0425 \ (X_1 + X_2)$$

Substituting for X_1 and X_2 from Equation 3.8 gives:

Total Aeration Basin SS Inventory =
$1.0425(0.4\ X_u + 0.25X_u) = 0.6776\ X_u$

Setting this equal to the total MLSS inventory of 6255 lb gives:

$$6225\ \text{lb} = 0.6776\ X_u$$

and

$$X_u = 9{,}230\ \text{mg/L}$$

X_1 can be calculated using Equation 3.6:

$$X_1 = 0.4\ X_u = (0.4)(9{,}230\ \text{mg/L})$$
$$= 3{,}690\ \text{mg/L}$$

X_2 can be calculated using Equation 3.8:

$$X_2 = 0.25X_u = (0.25)(9{,}230\ \text{mg/L})$$
$$= 2{,}310\ \text{mg/L}$$

The secondary clarifier SS loading rate for this operating condition would be:

$$G_a = (2{,}310\ \text{mg/L})(1.33\ \text{MGD})(8.34)/1{,}590\ \text{ft}^2$$
$$= 16.1\ \text{lb/ft}^2,\text{day}\ (78.6\ \text{kg/m}^2,\text{day})$$

This is plotted as point number 4 in Figure 47.

The benefits of two-pass step feed operation in allowing an activated sludge system to operate successfully with a poor settling sludge are illustrated in Figure 47 using the data from the previously calculated examples. Points number 1 through 3 are for plug flow operation with RAS flowrates of 0.33, 0.6, and 0.95 MGD (0.01, 0.026, and 0.04 m³/sec), respectively. Point number 4 is for two-pass step feed operation at an RAS flowrate of 0.33 MGD (0.01 m³/sec), as calculated above. These operating points show that an activated sludge with an SVI of approximately 220–230 mL/g requires an RAS flowrate of 0.95 MGD (0.04 m³/sec) when the aeration basin is operated in the plug flow mode and 0.33 MGD (0.01 m³/sec) when the aeration basin is operated in the two-pass step feed mode. This results both in lower RAS pumping costs and higher RAS, SS concentration (9230 mg/L vs 6300 mg/L) — which may reduce waste sludge processing and disposal costs.

This example illustrates the advantage of two-pass step feed over plug flow operation for dealing with bulking sludge. Operation in other step feed modes (such as three- or four-pass), contact stabilization, or step feed with RAS reaeration (a combination of step feed and contact stabilization) offer similar advantages. Indeed, any change which results in less dilution of RAS by influent wastewater will allow storage of activated sludge SS in the aeration basin and a decrease in the MLSS concentration in the secondary clarifier feed. For example, if the system depicted in Figure 46 had been placed in the contact stabilization mode, using the first pass for sludge stabilization and the second pass for contact (while maintaining the same MLSS inventory), the MLSS concentration in the secondary clarifier feed would have been reduced to 1,200 mg/L and the secondary clarifier SS loading rate would be 3.4 lb/ft²,day (41 kg/m²,day).

These examples have illustrated calculation procedures for determining the operating conditions when the mode of operation is changed. In some cases a simplified calculation procedure can be developed for a particular system operating under defined operating conditions, or operating diagrams can be developed for a particular system. Alternatively, these concepts can be used in a plant by changing plant operation and observing the results. The general principle that changes resulting in less dilution of RAS by influent wastewater will decrease the MLSS concentration of the secondary clarifier feed, can be used to guide changes in operating mode. After making changes, their impacts on SS distribution in the system can be monitored. In most cases redistribution of the SS takes less than one day, so that the results can be determined in a timely manner. The resulting secondary clarifier SS loadings can then be determined and the need for further changes assessed.

CHEMICAL ADDITION TO ENHANCE ACTIVATED SLUDGE SETTLING RATE

Several approaches to improving activated sludge settleability fall into this category in which the settleability of the sludge is improved without destruction or elimination of the filamentous organisms. Current practice in both municipal and industrial treatment plants is to use synthetic polymers to overcome the bridging or diffuse floc structure associated with excessive filamentous organism growth. Polymers also have been used to aid settling of activated sludges containing large amounts of water-retentive extracellular material (viscous or nonfilamentous bulking). When using organic polymers, no significant increase in sludge mass is experienced. Very wide variations in polymer type and dose have been reported. In general, either a high molecular weight, high cationic charge polymer alone or in combination with an anionic polymer is used. It is wise to perform jar testing frequently to estimate polymer dose, since this can change, and because overdosing can lead to deterioration in performance. An example of the results of such a jar test are presented in Figure 48. The data in Figure 48 suggest a polymer dose of about 2 mg/L would be appropriate for the sludge being tested. Polymers usually are added to the mixed liquor as it leaves the aeration basin or to the secondary clarifier center well. In general, the addition of polymers for bulking control is expensive compared to RAS chlorination— chemical costs in the range of $10–50/MG may be expected. Variations in the required polymer dose may occur with time due to a change in the nature of the MLSS and/or the accumulation of polymer in the mixed liquor. Consequently, the process must be carefully monitored and the required polymer dose reevaluated frequently.

In some instances inorganic coagulants/ precipitants such as ferric chloride and alum have been used to aid in activated sludge settling.

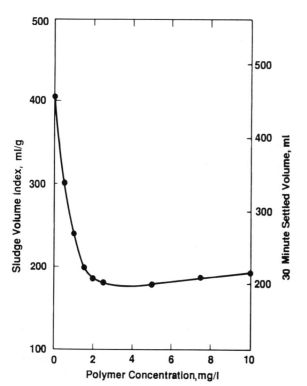

Figure 48. Effect of a cationic polymer on the SVI of a bulking sludge. (Singer et al., 1968). Reprinted by permission of the Water Environment Federation.

Reductions in SVI were noted by Eberhardt and Nesbitt (1968) and Zenz and Pivnicka (1969) when alum was added to activated sludge for the simultaneous precipitation of phosphate. These chemicals produce a precipitate that sweeps down the bulking activated sludge, increasing its bulk density and improving its settling rate. Finger (1973) used short term alum addition to the effluent end of the Seattle, Washington, Renton activated sludge plant to reduce SVI from about 130 mL/g to about 90 mL/g, increase sludge bulk density, and increase settling rates during periods of high (wet weather) influent flow. On one occasion, alum was added for 3 days at a dose of approximately 60 mg/L based on wastewater flow. This dose increased the total sludge inventory from 145,000 lb to 170,000 lb (a 17% increase), but because of a sludge density increase from 0.0075 g/mL to 0.0122 g/mL (a 63% increase), secondary clarifier sludge blanket levels were reduced. The beneficial effects of the 3 days of alum dosing were maintained for about 3 weeks.

Matsche (1982) reported several examples where the addition of ferrous sulfate to the influent was successful in controlling bulking. For example, at Neusiedl, Austria, the addition of ferrous sulfate (10–14 mg as Fe/L) to the influent of an activated sludge plant treating canning wastes reduced the SVI from 450 mL/g to 60–70 mL/g and caused the disappearance of type 021N, the filamentous organism responsible for the bulking.

Matsche (1982) also points out that activated sludge generated in plants without primary clarifiers settles better than those from plants with primary clarifiers, likely due to the presence of heavy primary SS in the activated sludge. Matsche (1982) also mentions the effect of dirt particles in sugar beet waste in improving activated sludge settleability. The PACT process, (Hutton and Robertaccio, 1975 and 1978) in which powdered activated carbon is added to activated sludge usually has good settling activated sludge, even in the presence of significant filamentous organism levels, because of the weighting effect of the carbon particles. The weighting action of inert biological SS has been employed in activated sludge process modifications such as the Hatfield and Kraus (Kraus, 1945) processes that recirculate anaerobic digester contents through the aeration basin system. While the addition of inorganic chemicals or inert biological solids can improve activated sludge settling rates, it imposes additional solids loads on the process. Moreover, the presence of inert solids cannot entirely offset the adverse impact of filamentous organisms on activated sludge settling.

SELECTIVE KILLING OF FILAMENTOUS ORGANISMS BY DISINFECTANT ADDITION

Two disinfectants, chlorine and hydrogen peroxide, have been used to selectively kill filamentous organisms and thus eliminate the symptoms of activated sludge bulking. Chlorine has been used in the form of a chlorine solution produced from a gas chlorinator and as sodium or calcium hypochlorite.

For this technique of bulking control, chlorination is our method of choice. The reasons for this are:

1. In the USA, and in terms of the relative amounts of chemical utilized for bulking control, chlorine is the more economical material.
2. In the USA, most municipal wastewater treatment plants disinfect secondary effluents with chlorine. Therefore, chlorine usually is available onsite, and the amount required for bulking control usually is quite small compared to the amount required for secondary effluent disinfection.

The recommended approach of using chlorination for bulking control can be tempered by specific circumstances. The most common of these are:

• Chlorine is neither available onsite nor can it be obtained immediately in sufficient quantities, and an extremely severe bulking problem requires immediate rectification.
• Chlorine is not used for effluent disinfection and there is a concern that chlorinated organic materials will be produced by chlorine addition to the system. This concern often is expressed by petrochemical and pulp and paper wastewater treatment plants. In one of the case histories presented later in this chapter, in-basin chlorination of a plastics manufacturing wastewater activated sludge showed no increase in effluent chlorinated organics during chlorination.

The proper application of chlorine for bulking control does not interfere with BOD_5 and SS removal efficiencies at the levels required for secondary treatment. As one of the case studies will show, there is always an increase (usually very small) in the effluent soluble COD, but not soluble BOD_5, during chlorination for bulking control. In one case where chlorination for bulking control was practiced on a laboratory scale, UCT biological nutrient removal system, nitrification, denitrification, and phosphorus release were not affected (Lakay et al. 1988). Phosphorus removal was reduced by high (8 kg $Cl_2/10^3$ kg MLSS,day) and prolonged (19 days) chlorine doses, but recovered rapidly (within 5 days) when the chlorination was stopped. In a pilot plant study of the Virginia Initiative Process (VIP) for enhanced biological phosphorus removal, RAS chlorination for bulking control with sodium hypochlorite at a Cl_2 dose of 2 kg $Cl_2/10^3$ kg MLSS,day for 5 days reduced SVI from approximately 270 mL/g to 200 mL/g (Daigger et al., 1988). Nitrification was not affected but the effluent total phosphorus concentrations increased to values close to those in the influent. Poor anaerobic zone phosphorus release and unusually high oxidation-reduction potential were observed during chlorination. Apparently, the Cl_2 interfered with the ability of the phosphorus-removing organisms to release phosphorus in the anaerobic zone, and this further inhibited phosphorus uptake in the aerobic zone. More typical anaerobic zone phosphorus release and ORP values and effluent phosphorus concentrations were reestablished almost immediately after chlorination was discontinued.

These results suggest that RAS chlorination can be used only sparingly as an operational tool for the control of sludge bulking in biological phosphorus removal plants.

Because we favor chlorination for bulking control we will deal first with this topic in detail and then present information on the use of hydrogen peroxide.

USE OF CHLORINATION FOR BULKING CONTROL

Prior to the 1970s, chlorination had been used occasionally for bulking control in the USA (Smith and Purdy, 1936; Tapleshay, 1945). Its use has become widespread in the USA since then. The capability to chlorinate RAS is now often designed into new treatment plants and frequently retrofitted into existing plants when they are expanded. Chlorination has been used widely throughout the world, including Germany (Frenzel and Sarfert, 1971; Frenzel, 1977; Bode, 1983), Austria (Matsche, 1977), Great Britain (Anon., 1983), and South Africa (Wiechers, 1983; Thirion, 1983). As previously indicated, its use is popular in the USA because the widespread use of chlorine for secondary effluent disinfection means that most municipal wastewater treatment plants have chlorine onsite. The quantities of chlorine required for bulking control usually are small compared with those required for secondary effluent disinfection, so that the question of there being sufficient chlorine supply available to perform both tasks usually does not arise. Rather, a common problem is to be able to control the existing chlorine delivery equipment at the low rates required for bulking control.

A variety of chlorine dosing points have been used (Figure 49). Without building additional mixing basins, chlorine can be added at 3 locations:

1. Directly into the aeration basin
2. Into an installed sidestream in which mixed liquor is pumped from and returned to the aeration basin
3. Into the RAS stream.

The most common and preferred dosing point is to the RAS stream. Other dosing points are used in wastewater treatment plants with long aeration basin hydraulic detention times, where dosing chlorine in the RAS would not provide a sufficient frequency of exposure of sludge inventory to chlorine, or when the RAS line is absolutely inaccessible.

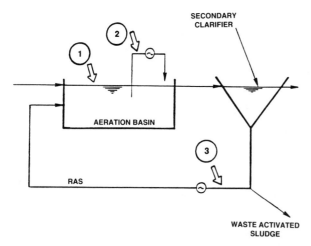

Figure 49. Chlorine dose points for bulking control.

Several criteria must be followed for successful bulking control by chlorination:

(a) A target value of the SVI (or some other activated sludge settling property) must be established.

The target SVI should be the value at, or below which, the plant, can be operated satisfactorily without any of the problems associated with poorly settling activated sludge. Not only should the operation of the secondary clarifier be considered in setting the value of the target SVI, but the impact of sludge settleability on solids processing units must also be evaluated.

(b) Chlorination should be used only when the target SVI (or other settling property value) is significantly and consistently exceeded. A daily (or more frequent) trend plot of SVI values should be made so that it can be seen when the target SVI is being approached. Such a trend plot will aid in making a decision on whether a given SVI value in excess of the target value is consistent with a trend, or is the result of a measurement error, or is an outlying data point. Trend plots also assist in making adjustments of chlorine dose to anticipate changes in SVI.

(c) Chlorine must be added in known and controlled doses to all of the activated sludge at a point of excellent mixing.

Ideally, a separate chlorinator should be dedicated to chlorination for bulking control because the chlorine dose rates required for bulking control usually are much lower than those employed for secondary effluent disinfection. When retrofitting an existing plant for chlorination for bulking control, an independently valved chlorine solution line can be taken off an effluent chlorinator to provide chlorine solution. It is imperative to accurately know the chlorine solution flowrate and concentration in this line. An independent rotameter in this chlorine solution line is mandatory.

As indicated above, the most common and favored chlorine application point is into the RAS line, although there are exceptions to this. If there is no common RAS line, the ability to chlorinate each individual RAS line must be provided. There must be sufficient valving and metering supplied to allow control and measurement of chlorine solution to each dosing point.

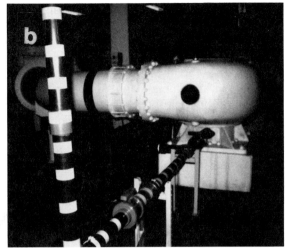

Figure 50. Examples of good chlorine dose points for bulking control: *a.* to the RAS line following an elbow; *b.* to the volute of a RAS pump; *c.* to the riser of an air lift RAS system; and *d.* to the "waterfall" of a traveling bridge clarifier RAS system.

Excellent initial mixing of chlorine solution and RAS is of the utmost importance. Reactions of chlorine in RAS are rapid so that, without excellent mixing, a large part of the RAS might not be contacted by the chlorine, while a small part could be overdosed. The result of poor initial mixing is the consumption of large amounts of chlorine without control of the bulking. Prolonged dosing at a point of poor mixing can lead to the production of turbid effluents (due to local over-chlorination) and a reduction in treatment efficiency (due to the excessive killing of floc-forming organisms). Examples of poor RAS chlorine injection points are RAS wet wells and the RAS or mixed liquor center wells of secondary clarifiers. The dose points of choice are at locations of high turbulence. Examples include: following elbows in pipes (Figure 50a), into the volute of RAS pumps (Figure 50b) or into their discharge, into and below the rising liquid level in the riser tube of an airlift RAS pump (Figure 50c), and through a manifold into the waterfall discharge to the aeration basin of an airlift/ return traveling bridge clarifier (Figure 50d). If no other choice is available, chlorine can be applied to RAS in an open channel; however, it should be at a point of turbulence (e.g., below a flow-measuring flume), and the chlorine injection should be multi-point through a pipe manifold or grid. Because such devices collect rags, the injection device should be designed so that it can be removed periodically for cleaning.

When retrofitting an activated sludge plant for RAS chlorination for bulking control, it is imperative to personally inspect the physical details of the system so that a suitable point(s) for chlorine injection can be specifically identified.

(d) Chlorine should be added at a point where the chlorine demand from the wastewater is at a minimum.

The purpose of chlorine addition for bulking control is disinfection (killing) of filamentous microorganisms. Any reactions with components of the wastewater that modify or reduce the dosed chlorine concentration will influence the way in which the chlorine kills the filamentous organisms. While the complexity of wastewater makes it impossible to identify each reaction of chlorine in it, some specific components are known to react with dosed chlorine. These are ammonia and reduced inorganics such as nitrite and sulfide. Jenkins et al. (1984) have shown that chlorine reacts rapidly and preferentially with ammonia to form monochloramine when dosed to RAS containing ammonia. Comparison of the chlorine residuals remaining after dosing activated sludge solids with chlorine in the presence of little, or excess ammonia (Cl_2/N ratio ≤ 5), and with an equal dose of monochloramine (Figure 51), shows that dosing activated sludge solids with chlorine in the presence of excess ammonia is equivalent to dosing with monochloramine. Monochloramine is much less potent than free chlorine as a disinfectant; however, it may remain available for a comparatively longer time. Thus the presence of ammonia does not materially affect the ability of chlorine dosed to the RAS to control bulking. The only exception is if the ratio of chlorine to ammonia-nitrogen at the chlorine dose point is such that the breakpoint reaction will occur (Cl_2/NH_3-N ratio of > 10:1). In these circumstances, the chlorine will oxidize the ammonia-nitrogen to nitrogen gas, thereby preventing it from attacking the filamentous organisms. The conditions necessary for the breakpoint

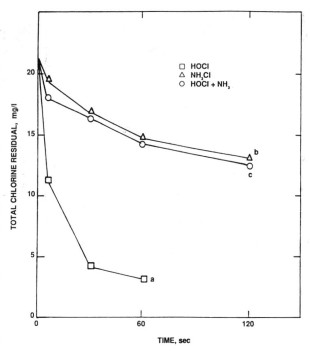

Figure 51. Chlorine residual in RAS as a function of initial chlorine species, reaction time and the presence of ammonia (initial chlorine dose = 21 mg Cl_2/L; ammonia concentration = 0.8 mg NH_3-N/L for *a.* and *b.* and 11.3 mg NH_3-N/L for *c.*; SS = 2.77 g/L) (Neethling et al., 1982).

reaction occur only infrequently and can be overcome by increasing the chlorine dose.

The only other common side reaction of chlorine that can reduce its effectiveness in killing filamentous organisms for bulking control is with nitrite. Chlorine reacts rapidly and stoichiometrically with nitrite, which can be present in activated sludge from partial nitrification or sometimes from denitrification. The reaction:

$$HOCl + NO_2^- \rightarrow NO_3^- + HCl$$

consumes 5.1 mg Cl_2/mg NO_2^- – N. It takes place rapidly, consumes chlorine before it has a chance to react with filamentous organisms, and thus impairs the ability of chlorine to control bulking.

It is very unusual to encounter significant levels of sulfide in RAS, so that the consumption of chlorine by the oxidation of sulfide is not usually significant in RAS chlorination for bulking control. We have encountered sulfide interference with RAS chlorination once when a broken secondary clarifier sludge scraper allowed sludge accumulation to a point where the sulfide levels consumed significant amounts of chlorine (and caused *Thiothrix* spp. bulking).

Chlorine reacts with the dissolved and particulate organic matter present in wastewater. The nature, rate, and extent of these reactions has not been quantified further than to say it takes place at rates slower than the reactions of chlorine with ammonia and nitrite. Addition of chlorine to mixtures of RAS and influent wastewater, such as might exist close to the head-end of the aeration basin, should be avoided because chlorine will be consumed by reaction with the high concentrations of organic matter and possibly sulfide, and thus will not be available for killing of filamentous organisms. High chlorine doses will be required for bulking control under these conditions.

Addition of chlorine to the incoming wastewater prior to the point where it mixes with RAS is completely ineffective for filamentous organism disinfection and bulking control.

(e) Chlorine doses are measured conveniently on the basis of the sludge inventory in the plant (the overall chlorine mass dose in kg $Cl_2/10^3$ kg SS inventory,day). The chlorine concentration at the dose point (mg/L), and the frequency of exposure of the solids inventory to chlorine (times/day) must be considered in choosing a dose point.

The most commonly used chlorine dose parameters are dose concentration, overall mass dose rate, and frequency of exposure of solids inventory to chlorine dose (Beebe and Jenkins, 1981; Beebe et al., 1982; Jenkins et al., 1984).

All of these parameters are expressed in terms of chlorine dose rather than chlorine residual. It is the nature and concentration of the chlorine residual that is the important parameter in the disinfection of filamentous organisms. However, because of the complexity and variability of activated sludge composition, it is extremely difficult (if not impossible) to predict chlorine residual. Therefore, chlorine dose must be employed.

This, and the differences in susceptibility of various filamentous organisms to chlorine, are reasons why significant variation in the values of dosing parameters for successful bulking control exist from one plant to another and even from one time to another at the same plant (Neethling et al. 1985a; 1985b).

Overall Mass Dose (kgCl$_2$/10^{3}kgSS, day) is a convenient practical dosing parameter because it tells the operator how much chlorine has to be dosed in a day and allows the dosing equipment to be adjusted accordingly. It also is the fundamental dosing parameter from the standpoint of the process objective—to kill a fraction of the activated sludge inventory each day. Daily organism kill is proportional to the daily mass dose of chlorine (kg/day), while the system SS inventory is proportional to the total mass of organisms in the system.

From a practical standpoint, when designing a chlorination system the overall mass dose is computed to allow sizing of the chlorination system. In computing chlorine dose based on the sludge inventory (kgCl$_2$/10^{3}kg SS, day), the sludge SS inventory in the secondary clarifiers should be included. Where "two basin" processes such as contact stabilization or sludge reaeration are employed, the sludge SS inventory in both basins and in the secondary clarifiers should be included in calculating the overall chlorine mass dose.

In most municipal wastewater treatment plants the RAS line is the best place to dose chlorine. For a chlorination dose point on the RAS line, the minimum frequency of exposure should be determined. A rule of thumb is that if the frequency of exposure is ≥ 3 times per day, then the selected dose point will provide a sufficient frequency of exposure of SS inventory to chlorine. In determining frequency of exposure, account should be taken of the maximum rate of sludge recycle available. It may be possible to obtain the desired frequency of exposure by increasing sludge recycle rates to values higher than those that are normally used. Some RAS chlorination systems (Bode, 1983) have operated well at frequencies as low as 1.4 per day. Neethling et al. (1985a) showed that the required frequency of exposure can be related to the relative growth rates and efficiencies of kill of filamentous and floc-forming organisms. Simply put, the filamentous organisms must be killed faster than they grow in the aeration basin for chlorination to be successful.

In activated sludge plants with long aeration basin hydraulic detention times and large SS inventories, the daily SS flux in the RAS line is a small fraction of the total solids inventory. Because of this, the frequency of exposure of SS inventory to a chlorine dose point located on the RAS line will be very low. Figure 52 shows the effect of employing typical overall mass dose rates of chlorine for bulking control at a plant with a typical domestic wastewater treatment configuration, and at a typical high-strength industrial wastewater treatment plant with a 4-

a. "DOMESTIC WASTEWATER" CASE

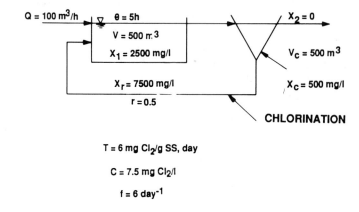

$Q = 100\ m^3/h$ $\theta = 5h$

$V = 500\ m^3$

$X_1 = 2500\ mg/l$

$X_r = 7500\ mg/l$

$r = 0.5$

$X_2 = 0$

$V_c = 500\ m^3$

$X_c = 500\ mg/l$

CHLORINATION

$T = 6\ mg\ Cl_2/g\ SS,\ day$

$C = 7.5\ mg\ Cl_2/l$

$f = 6\ day^{-1}$

b. " INDUSTRIAL WASTEWATER" [LONG AERATION TIME] CASE

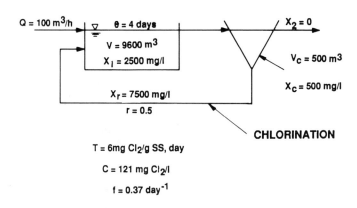

$Q = 100\ m^3/h$ $\theta = 4\ days$

$V = 9600\ m^3$

$X_l = 2500\ mg/l$

$X_r = 7500\ mg/l$

$r = 0.5$

$X_2 = 0$

$V_c = 500\ m^3$

$X_c = 500\ mg/l$

CHLORINATION

$T = 6mg\ Cl_2/g\ SS,\ day$

$C = 121\ mg\ Cl_2/l$

$f = 0.37\ day^{-1}$

Figure 52. Chlorination parameters for a typical domestic wastewater treatment plant (*a.*) and a long aeration time treatment plant (*b.*). T = overall Cl_2 mass dose rate, C = Cl_2 concentration at dose point and f = exposure frequency of sludge inventory to Cl_2.

day hydraulic detention time aeration basin (industrial wastewater case).

In Figure 52a, an activated sludge plant with an aeration basin mean hydraulic detention time of 5 hr and an RAS rate of 50 m^3/hr (r = 0.5) is chlorinated in the RAS stream at a dose of 6 kg $Cl_2/10^3kg$ MLSS, day. The chlorine dose concentration is 7.5 mg Cl_2/L. The SS inventory passes the chlorine dose point 6 times per day. These conditions of chlorine concentration and frequency of solids inventory exposure to chlorine are satisfactory for bulking control. In Figure

52b, an activated sludge plant with an aeration basin having a mean hydraulic residence time of 4 days and an RAS rate of 50 m^3/hr (r = 0.5) is chlorinated in the RAS stream at a dose of 6 kg $Cl_2/10^3kg$ MLSS, day. The chlorine dose is 121 mg Cl_2/L. The SS inventory passes the chlorine dose point once every 2.7 days, or with a frequency of 0.37 times per day. These conditions of chlorine concentration and frequency of SS inventory exposure to chlorine are not satisfactory for bulking control. A small part of the SS inventory is exposed to excessively high chlorine doses. The exposed sludge flocs would be destroyed, resulting in high effluent turbidity; the sludge as a whole would not be exposed frequently enough to the chlorine dose to control filamentous organism growth. Thus, bulking would continue and, in addition, a turbid effluent would be produced. In such cases, a chlorination system using several dose points placed directly in the aeration basin (Figure 49) should be installed.

(f) Reliable measurements of sludge settling, effluent quality and sludge quality are required when chlorination is used for bulking control.

The use of chlorine for bulking control entails the purposeful addition of a disinfectant to a biological system. Because of this, control tests to assess the effects of chlorine addition must be performed frequently and carefully. Such control tests should include the following:

1. A reliable and appropriate measurement of sludge settling—such as DSVI, stirred SVI, zone settling rate, or sludge blanket depth in the secondary clarifiers at known wastewater and RAS flowrates.
2. A measure of secondary effluent turbidity—increases in secondary effluent turbidity accompanied by rapid decreases in SVI indicate chlorine overdose. Increased secondary effluent turbidity indicates that significant amounts of activated sludge floc material are being destroyed by chlorine. Indeed, one of the observations commonly made when chlorine is applied (improperly) in shock doses for bulking control is that the secondary effluent turns "milky." This is due to the presence of small particles of destroyed activated sludge floc produced by excessively high chlorine concentrations. Turbidity can be measured with a turbidimeter or can be appraised visually by making transparency measurements directly in the secondary clarifier or chlorine contact chamber with a Secchi disc. (This latter measurement is possible only when sludge blanket levels are lower than the extinction depth of the Secchi disc!).
3. Microscopic examination of the activated sludge—this type of examination should be conducted routinely; it is very important

when chlorination is being used for bulking control because the effects of chlorine on both the filamentous organisms and on the flocs are readily observable. When chlorination is beginning to have an effect on filamentous organisms it often is possible to see progressively deformed cells in the filaments.

The progressive effect of chlorination on type 1701 and *Thiothrix* spp. is illustrated in Figure 53. Both organisms have sheathed trichomes. Chlorine progressively causes cells to deform and to detach from the sheath and "ball-up"; gaps then appear in the sheaths left by the disappearance of cells; finally, empty and broken sheaths remain. With *Thiothrix* spp., this chain of events can be preceded by the disappearance of intracellular sulfur granules, which presumably are oxidized by the chlorine.

The initial effects of chlorine on filamentous organisms described above can precede observable improvements in activated sludge settling properties, and therefore can be used as an early signal that chlorination is beginning to exert control over filamentous organism growth. These observations enable a plant microbiologist to tell the plant operator that a satisfactory chlorine dose has been reached.

Overdose of chlorine also can be detected microscopically. The total elimination of filamentous organisms and the presence of small, broken-up flocs together with fine particles are signs of chlorine overdosing. Another sign of chlorine overdose is the complete loss of protozoa and rotifers. If these are present and active, then chlorine is not being overdosed.

Chlorination has been used successfully to control bulking caused by many types of filamentous organisms (Table 25). We have had universal success with it. Chlorination appears to be effective against filamentous organisms growing both inside and extended outside of the activated sludge flocs. The filamentous organisms found in activated sludge seem to vary in their susceptibility to chlorine, with *Thiothrix* spp. and type 021N being quite susceptible and *M. parvicella* and *Nostocoida* spp. being rather resistant, and the remainder falling somewhere in between. There is only one report (Frenzel and Sarfert, 1971) in which it is claimed that one type of filamentous organism was not affected by chlorination.

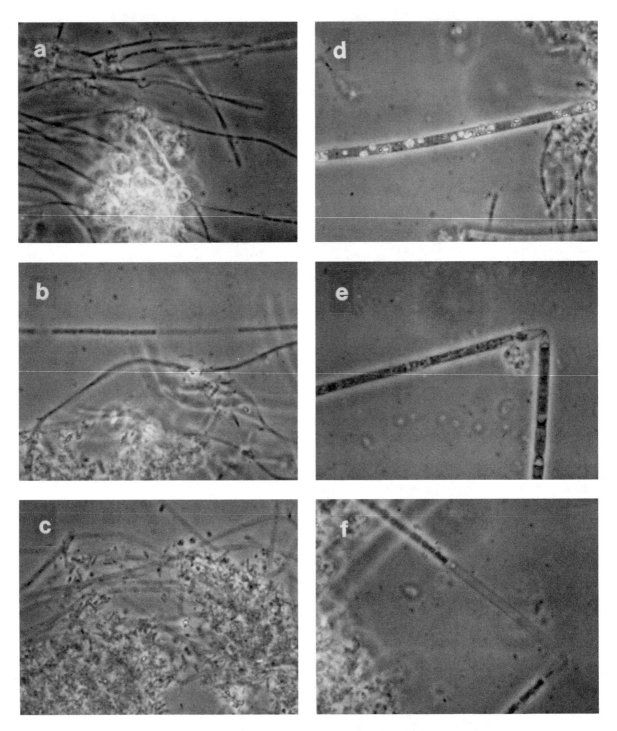

Figure 53. Progressive effect of chlorination on type 1701 (*a.*, *b.* and *c.*) and *Thiothrix* spp. (*d.*, *e.* and *f.*): type 1701. *a.* and *d.* no chlorination; *b.* and *e.* moderate chlorination effects; and *c.* and *f.* severe chlorination effects (all 1000X phase contrast).

Table 25. Bulking Filamentous Organisms That Have Been Successfully Controlled by Chlorination (Partial List Only)

Location	Major Filamentous Organism(s) Causing Bulking
City of Albany, GA	type 0041 type 0092 *H. hydrossis* type 1701
Malibu Mesa, CA	type 1701 type 021N
City of Los Angeles Terminal Island, CA	type 021N type 0581
Carrousel pilot plant treating brewery wastewater	type 0041 type 0092
City of Pacifica, CA	*S. natans* type 1701 type 1863
City of San Jose, CA	*H. hydrossis* type 1701 type 021N type 0803 type 0041 type 0675
Derry Township Municipal Authority Hershey, PA	*Thiothrix* spp. *M. parvicella*
Oro Loma Sanitary District, CA	*H. hydrossis* type 1701
Weyerhaueser Longview, WA (NaOCl)	type 0675 type 1851
Miller Brewing Co. Fulton, NY	type 0041 type 0675
Stroh Brewing Co. Longview, TX (NaOCl to Aeration Basin)	type 1851 type 021N *N. limicola II*
Champion International (Pulp and paper) Courtland, AL	*Thiothrix* spp. type 1701
Monterey Regional Water Pollution Control Plant, Monterey, CA	*N. limicola II* type 1701 *Thiothrix* spp.
City of Redding, CA	type 1701 *S. natans*
Westchester County Yonkers, NY	type 1701 *N. limicola II*
Thames Water Authority (UK) (NaOCl)	type 021N
Central Contra Costa Sanitary District Concord, CA	type 021N *M. parvicella* *Thiothrix* spp.

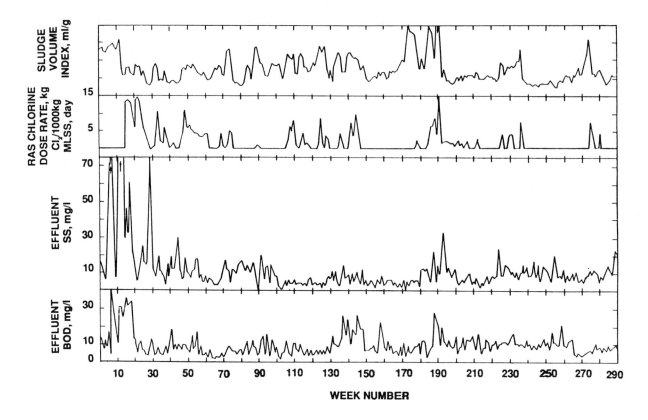

Figure 54. Control of bulking by RAS chlorination at Albany, GA wastewater treatment plant (Jenkins et al., 1984). Reprinted by permission of the Institution of Water and Environmental Management.

CASE HISTORIES OF BULKING CONTROL USING CHLORINATION

Several case histories will be given to illustrate the use of chlorination for bulking control. These include two municipal wastewater treatment plant case histories illustrating the successful long-term use of RAS chlorination, and two industrial wastewater treatment plant case histories illustrating the use of chlorine directly in the aeration basin.

City of Albany, Georgia

The city of Albany, Georgia, USA, is a 20 MGD (0.88 m³/sec) (design) activated sludge plant which, at the time these data were collected, received 13–16 MGD (0.57–0.7 m³/sec) of a wastewater in which approximately 50% of the BOD_5 and SS loads were from industry (paper processing and convenience foods manufacturing). The plant experienced bulking problems to such a degree that with only approximately 12 MGD (0.5 m³/sec) of influent flow, attainment

of secondary effluent criteria (30 mg/L monthly average for BOD_5 and SS) was not possible.

At Albany, the permanent RAS chlorination system injects chlorine solution into a series of well-mixed RAS wet wells. The installation consists of a Wallace and Tiernan Chlorinator with a 2200 lb/day (1000 kg/day) capacity, and a 2-in.(50 mm) injector with 2-in.(50 mm) PVC injector water and chlorine solution lines. The chlorine injection system is enclosed in a sheet metal housing for protection. The chlorine solution line is a 2-in.(50 mm) PVC line leading to a 16-ft (5 m) long, 2-in.(50 mm) PVC header with 1/4 to 3/8-in.(10–15 mm) diameter holes at 4-in.(100 mm) spacing. The system was constructed in November 1977 for $4,953.

Figure 54 presents over six years of data concerned with RAS chlorination at Albany, Georgia. When chlorination for bulking control was initiated, the RAS chlorination system did not exist. Because of the urgency to control SVI, chlorine was added to a wet well where the entire RAS stream mixed with primary effluent. This mode of chlorination was commenced on Week

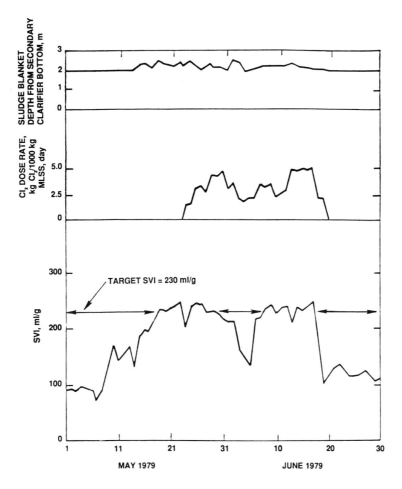

Figure 55. Use of target SVI to control RAS chlorination dose for bulking control at the Albany, GA wastewater treatment plant. (Jenkins et al., 1982).

15 and continued until Week 26, when the change to RAS chlorination was made. Chlorine was being added to the mixture of RAS and primary effluent, and doses of 5–15 kg $Cl_2/10^3$kg SS, day were needed to control bulking. Immediately after the change from chlorinating the mixture of RAS and primary effluent to chlorinating RAS alone, chlorine doses that satisfactorily controlled bulking for the mixture of RAS and primary effluent were "overdoses" for RAS alone. Figure 54 shows that even though the chlorine dose declined during the changeover of the chlorine dose point (from 4.7 to 2.9 kg $Cl_2/10^3$kg SS, day), there was an increase in secondary effluent SS (and to a lesser extent in secondary effluent BOD_5) that resulted from floc breakup caused by the chlorine overdose.

At various times during the period covered by the data in Figure 54 the target SVI was changed, and in some instances (Figure 54, Weeks 180–200) RAS chlorination was not initiated soon enough to prevent secondary effluent deterioration by a loss of sludge blanket solids in the effluent during peak flow periods. To illustrate the chlorine dosing technique for bulking control as employed by the city of Albany, a specific bulking incident is examined in detail in Figure 55. At this time, the target SVI was 230 mL/g. The chlorine dose was started at low levels and gradually increased until the SVI responded by stabilizing and then falling. In the middle of this period (May 30–June 6), the SVI fell to below the target value. The RAS chlorine dose was reduced in response to this. On 18–20 June the SVI dropped to approximately 100–130 mL/g. The chlorine dose to the RAS was reduced and then turned off. RAS chlorination for bulking control continues to be used in the Albany plant.

City of San Jose/Santa Clara Water Pollution Control Plant (SJ/SC WPCP), California

At the time these data were collected, the SJ/SC WPCP provided tertiary treatment to 100 MGD (4.4 m³/sec) of municipal wastewater. During Jul-Sep (the peak load season) flows increased approximately 20%, and BOD_5 loading approximately doubled due to cannery waste discharges. Effluent discharge criteria at that time included 30-day average BOD_5 and SS of 10 mg/L each and receiving water undissociated ammonia standards that dictated virtually complete nitrification. The plant has two stages of activated sludge with tertiary effluent filtration. It had plant upsets due to bulking in the secondary activated sludge system during the peak load seasons of 1979 and 1980. Interim measures including RAS chlorination in both the secondary and tertiary activated sludge systems, and provision of supplemental oxygen and ammonia, were used to prevent bulking in the 1981 and 1982 peak load seasons (Beebe et al., 1982).

In the first-stage activated sludge system, the initial chlorine injection point to the RAS was through a PVC pipe with a 4-ft (1.2 m) freefall into a 20 ft × 16 ft × 13 ft (6m × 5m × 4m) deep wet well. It soon became evident that this arrangement was ineffective and chlorine could be smelled in the vicinity. Extension of the 2-in.(50 mm) PVC pipe to a point 5 ft (1.5m) below the RAS surface in this wet well virtually eliminated loss of gaseous chlorine, but SVI control was still difficult. Doses of 8 to 10 kg $Cl_2/10^3$kg SS, day had little effect on SVI, but created a turbid secondary effluent.

Following these poor experiences, two alternate chlorination devices were installed:

- diffusers about mid-depth across the full width of the aerated mixed liquor channels leading from the aeration basins to the secondary clarifiers
- single outlet injectors through the clean-out ports about 6 in.(15 cm) upstream of closed-impeller, low-speed centrifugal RAS pumps (Figure 50b).

The data presented in Figure 56 show that the pump clean-out port injection point, with its superior mixing, provided more rapid, predictable and efficient SVI control. Operating data

indicated that approximately 20 times more chlorine was added when chlorine was dosed to the RAS well rather than to the RAS pump clean-out port. As a result of this success a similar chlorine injection system was installed in the second-stage activated sludge system. It, too, has been used successfully for bulking control without compromising the nitrifying ability of the second stage activated sludge system. Injection of chlorine solution into the first- and second-stage activated sludge system RAS pumps for over nine years has had no deleterious effects on these pumps, even though the second-stage RAS pumps have brass bearings. Impellers in all RAS pumps are cast iron, and great care is taken to make certain that chlorine solution is never fed to an out-of-service pump.

RAS chlorination for control of filamentous bulking has been developed to a high degree at the SJ/SC WPCP (Figure 57). Chlorine doses are regulated routinely on the basis of a target SVI value with input from the plant microbiologist, who observes activated sludge samples on a daily basis during periods of RAS chlorination. During the peak load season of 1981, extremely low target SVI values (60 to 80 mL/g) were used in the first-stage activated sludge system, so that high solids loading rates (30–50 lb SS/ft², day)(150–250 kg SS/m², day) could be applied to the secondary clarifiers. To maintain these low SVI values, chlorine doses to the first-stage activated sludge system often were high—in the range from 8 to 16 kg $Cl_2/10^3$kg VSS, day. These high chlorine doses produced significant floc destruction, resulted in turbid secondary effluents, and caused increases in secondary effluent dissolved total organic carbon (TOC) concentrations. The elevated turbidity and the TOC increase from approximately 15 to 35 mg/L could be tolerated at the SJ/SC WPCP because the downstream second-stage activated sludge system readily polished the effluent. Applying such high chlorine doses and obtaining a high-quality secondary effluent would be difficult if the first-stage activated sludge plant were operated without the second-stage activated sludge system to capture the fine solids.

RAS chlorination has also been effective for bulking control in the second stage, nitrifying activated sludge system (Figure 58). At the chlorine dosage used during the 1981 peak load season (2–4 kg $Cl_2/10^3$kg VSS, day; 1.5 to 3.0 mg Cl_2/L dose in the RAS stream), nitrification efficiency was unaffected. This observation is consistent with that of Strom and Finstein

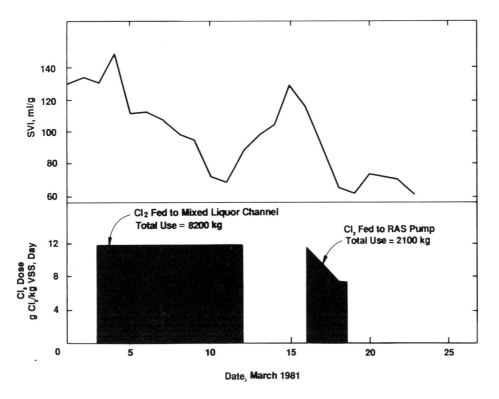

Figure 56. Comparison of RAS chlorination effectiveness when dosing chlorine to either the mixed liquor channel or the RAS pump. (Beebe et al., 1982).

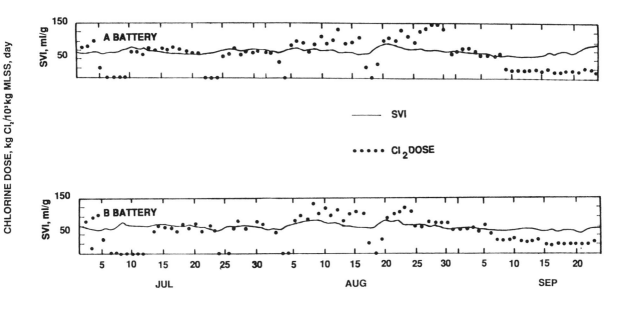

Figure 57. SVI and chlorine dose to RAS in the two first-stage activated sludge systems at the San Jose/Santa Clara, CA water pollution control plant – 1981 peak load season. (Beebe et al., 1982).

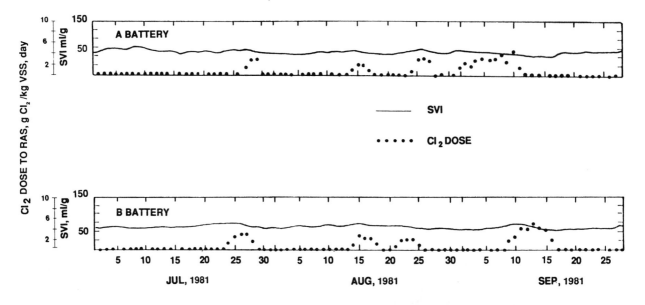

Figure 58. SVI and chlorine dose to RAS in the two second-stage activated sludge systems at the San Jose/Santa Clara, CA water pollution control plant—1981 peak load season. (Beebe et al., 1982).

(1977). In the nitrifying system the decrease in SVI with commencement of RAS chlorination was much faster than it was in the first-stage activated sludge system. Whether this was due to the presence of the more germicidal free chlorine in the nitrifying system, to differences in the types of filamentous organisms causing bulking (type 0041 in the second stage, type 1701 in the first stage), or to differences in system sludge growth rates is not known. Chlorination of the RAS continues in use at the SJ/SCWPCP.

Stroh Brewing Co., Longview, Texas
[Chlorination Directly into the Aeration Basin (Campbell et al., 1985)].

At the time these data were collected, the Stroh Brewing Company operated a brewery and container manufacturing facility in Longview, Texas. The combined wastewater flows from these plants are treated by an extended aeration activated sludge plant. At the time these data were collected, combined wastewater flow averaged 1.75 MGD (0.08 m³/sec) and was comprised of 1.6 MGD (0.07 m³/sec) of brewery wastewater and 0.15 MGD (0.01 m³/sec) of a wastewater from container manufacturing. Typical wastewater characteristics were BOD_5 = 1500 mg/L, COD = 2500 mg/L, TSS = 800 mg/L, TKN = 7 mg/L, and total P = 25 mg/L.

The treatment facilities consisted of a mechanically cleaned bar screen, grit removal, nitrogen (ammonia) addition, two parallel completely-mixed aeration basins and secondary clarifiers, a chlorine contact chamber, and post aeration.

The Longview plant had always been subject to filamentous bulking problems, which were most severe during the spring and summer (Apr-Oct). The causative filamentous organisms usually were types 1701 and 021N. The poor settleability that resulted from the presence of large amounts of these filamentous organisms required the use of significant quantities of cationic polymer to produce acceptable sludge separation. In an effort to reduce the level of filamentous organisms and, consequently, improve sludge settleability and reduce polymer costs, chlorination of the RAS was practiced during 1979. Chlorine was fed to the RAS at a rate of 3 kg Cl_2/10^3 kg MLSS, day—the maximum rate that could be achieved with the existing chlorine dosing equipment. Throughout the 19-week period that chlorine was dosed, filamentous organism growth was apparently unaffected, and control of sludge bulking was ineffective (Figure 59). It was postulated that the inability to control bulking by RAS chlorination was due to the infrequent exposure of activated sludge solids to chlorine afforded by the chlorine dose point on the RAS line. At the Longview plant, where aeration basin average hydraulic residence time typically is 4 days, the

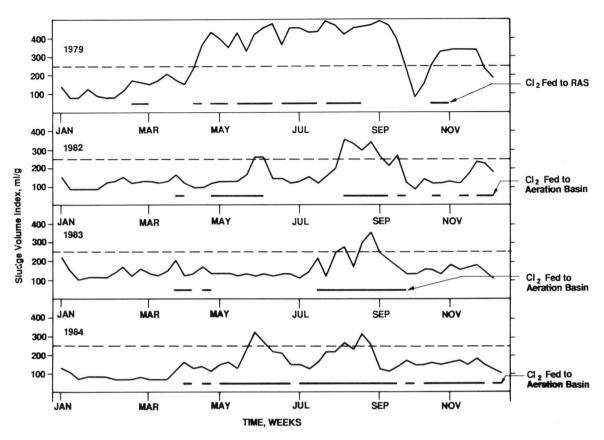

Figure 59. Weekly average SVI during 1979–84 at Stroh Brewing Co., Longview, TX activated sludge plant. (Campbell et al., 1985).

Table 26. Annual Usage and Cost of Bulking Control Chemicals at Stroh Brewing Co., Longview, TX Activated Sludge Plant

	1979	1980	1981	1982	1983	1984
Average BOD$_5$ Load, 10^3 lb/day	22,000	20,000	23,000	23,000	20,000	23,000
Chlorine Dose Point	RAS	—	—	Aeration Basin	Aeration Basin	Aeration Basin
Polymer Consumption, gal/yr	7,600	2,160	2,360	1,090	213	182
Chlorine Consumption, 10^3 lb/yr	30	0	0	43	27	61
Polymer Cost, $/yr	63,000	18,000	19,700	9,100	1,800	1,500
Chlorine Cost, $/yr	3,600	0	0	5,200	3,200	7,300
Total Cost, $/yr	66,600	18,000	19,700	14,300	5,000	8,800
Average SVI, mL/g	285	286	164	160	159	156

activated sludge inventory passes a chlorine dose point on the RAS line approximately 0.4 times per day.

This situation was rectified in 1982 by relocating the chlorine dose point from the RAS line to two points in each aeration basin (for a total of four dose points) located close to four of the six surface aerators and 6 ft (2m) below the liquid surface. Chlorine solution at a dose rate of 3 kg $Cl_2/10^3$kg MLSS, day was split evenly between the four dose points. Figure 59 shows the results of using this in-basin chlorination system for bulking control during the period 1982 to 1984. The total costs for bulking control chemicals (polymer plus chlorine) have been reduced from $66,600/yr to $8,800/yr by use of the in-basin chlorination system, while the BOD_5 load has remained approximately constant (Table 26).

Plastics Manufacturing Wastewater Activated Sludge System, West Virginia
[In-basin Chlorination for Bulking Control (Campbell et al., 1985)].

This example is given to illustrate the effects of a chlorine overdose on activated sludge and to show how rapidly a system recovers. The conventional activated sludge system treated 1.4 to 3.0 MGD of wastewater from a complex plastics manufacturing site. Typical waste characteristics were COD = 1200–1800 mg/L, BOD_5 = 700–1100 mg/L, and TSS = 30–100 mg/L. Typical F/M values were 0.2 to 0.5 kg COD/kg MLSS,day with MLSS levels between 3000 and 5000 mg/L. Activated sludge settleability at the plant was related to MLSS concentration—as the MLSS concentration increased, so did the SVI. At high SVI (e.g., >120 mL/g), two operating problems occurred. First, the clarifier sludge blanket levels increased, which often resulted in high effluent SS levels. Second, the sludge dewatered poorly in the centrifuges, reducing the system solids removal capacity. When this happened, the MLSS concentration increased to cause further increases in SVI, and so on.

The major filamentous organisms causing bulking were type 1702, and to a lesser extent, *N. limicola*. Type 1702 grew largely inside the flocs, stretching them out and making them diffuse. Type 1702 is related to type 1701 (Eikelboom, 1977), the growth of which in activated sludge is caused by a DO concentration too low for the applied F/M (Richard et al., 1985).

The in-basin chlorine dosing system employed chlorine solution from a gas chlorinator and four dose points in the aeration basin. In-basin chlorine addition was first tested on Mar 9, 1984, when the SVI exceeded 120 mL/g and solids settling was significantly impaired. Figure 60 depicts the Cl_2 addition rates and SVI values throughout the test period. Thirty-minute settleable solids were monitored every 4 hr for control purposes and SVI, which was calculated twice each day, was used in the after-the-fact analysis of the data. An initial dose rate of 3 kg $Cl_2/10^3$ kg MLSS,day was applied for 3 days with no effect on the SVI—in fact, the SVI continued to rise. On Mar 12, 1984, the Cl_2 dose rate was increased to 5 kg $Cl_2/10^3$ kg MLSS,day and held at this level for about 5 days. Again, no noticeable improvement in sludge settleability occurred. On Mar 16, 1984, the Cl_2 dose rate was increased to 6.7 kg $Cl_2/10^3$ kg MLSS,day, and after being maintained at this rate for approximately 3 days, the SVI started to decrease. On Mar 19, 1984, the Cl_2 dose rate was reduced to and maintained at 4 kg $Cl_2/10^3$ kg MLSS,day for 3 days. The SVI continued to decrease slowly, but then leveled off at approximately 120 mL/g. On Mar 22, 1984, the Cl_2 dose rate was increased to 8 kg $Cl_2/10^3$ kg MLSS,day to test the impact of a higher dose on SVI levels. After a little over 3 days at this Cl_2 dose rate, the SVI decreased rapidly, and the Cl_2 dose rate was then decreased to 5 kg $Cl_2/10^3$ kg,day for 2 days and stopped on Mar 27, 1984. The SVI began to increase before leveling off at 90 mL/g.

Microscopic examination of the activated sludge during in-basin chlorination revealed that SVI reduction by Cl_2 was accompanied by both significant filament and floc damage. As the Cl_2 dose was increased, type 1702 filaments successively showed deformed cells, empty spaces where cells had lysed, completely empty sheaths, and broken sheaths. Flocs showed progressively greater breakup with production of fine suspended particles. Effects such as these are indicative of very high Cl_2 doses (overdoses). It is likely that such high Cl_2 doses were required because type 1702 growth was largely within the floc; thus, significant floc destruction was required for the Cl_2 to get at the type 1702 filament.

While the in-basin chlorination was judged to be successful, the results demonstrated that high in-basin Cl_2 doses adversely affected effluent TSS (Figure 61), soluble COD (Figure 62), and turbidity levels. For example, following peak Cl_2 dose rates, the effluent TSS concentration increased to 60–80 mg/L from the typical 30–50 mg/L range and effluent soluble COD increased from normal levels of 40–80 mg/L to levels as high as 150

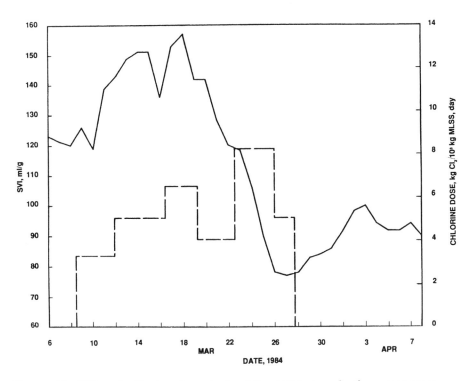

Figure 60. SVI and chlorine dose during Mar 1984 at a plastics wastewater activated sludge plant. (Campbell et al., 1985).

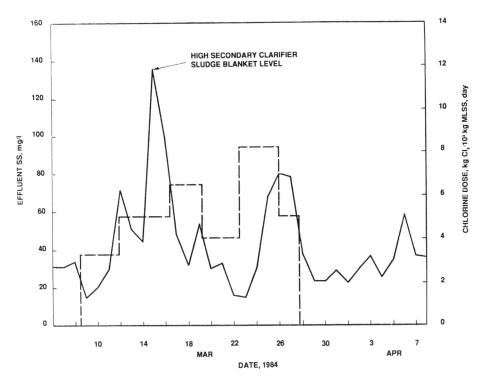

Figure 61. Effluent TSS concentration and chlorine dose during Mar 1984 at a plastics wastewater activated sludge plant. (Campbell et al., 1985).

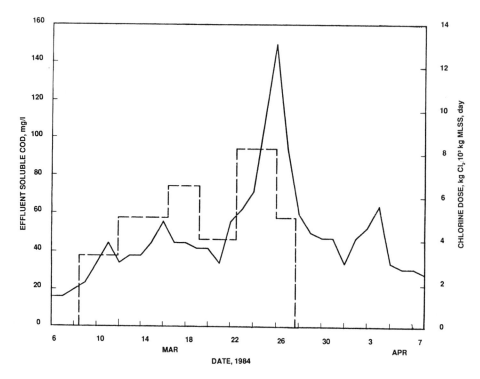

Figure 62. Effluent soluble COD concentrations and chlorine dose during Mar 1984 at a plastics wastewater activated sludge plant. (Campbell et al., 1985).

mg/L. Microscopic examination of the mixed liquor showed that rotifers were active, indicating a healthy microorganism population. Also, oxygen uptake rates remained at or above normal levels. Thus, it appears the increase in effluent soluble COD was due to the release of soluble cell material as the cell walls of dead organisms ruptured, and not due to a decrease in microorganism activity. These impacts, however, illustrated the need for very careful control of Cl_2 dose rate and duration during bulking control.

A second in-basin chlorination treatment was performed during a further bulking incident that occurred in October 1984. The results of this treatment were very similar to the first in that the filamentous organisms were successfully controlled with Cl_2 dosages of 5.5 to 6.5 kg $Cl_2/10^3$ kg, day. SVI was reduced from approximately 180 mL/g to approximately 100 mL/g and modest increases in effluent SS and COD were observed.

Effluent samples were analyzed for chlorinated organics to determine whether such compounds were formed during Cl_2 addition. With the exception of methylene chloride, other chlorinated organic priority pollutants were present at concentrations near or below the analytical detection

limit. Methylene chloride levels were less than 100 μg/L and within the normal variability of routine plant operation. On the basis of these results, the generation of chlorinated organic compounds during in-basin Cl_2 addition for bulking control was concluded to be insignificant. Similar conclusions have been made by van Leeuwen and van Rossum (1990) when using in-basin chlorination of Daspoort, Pretoria, South Africa activated sludge in laboratory-scale experiments. No detectable organohalogens were found in the secondary effluent when employing in-basin chlorination at a rate of 8 kg$Cl_2/10^3$ kg MLSS,day in this nutrient removal activated sludge plant.

USE OF HYDROGEN PEROXIDE FOR BULKING CONTROL

Hydrogen peroxide (H_2O_2) has been used for bulking control in a fashion similar to chlorine. H_2O_2 doses and length of application for effective filamentous organism reduction and bulking control vary from case to case (Table 27). In general, it can be deduced from Table 27 that the effective H_2O_2 doses are somewhat higher than would be required for chlorine. However, since no compar-

Table 27. Use of Hydrogen Peroxide for Bulking Control

Location	Successful H_2O_2 Concentration, mg H_2O_2/L	Reference
Petaluma, CA full-scale plant	9–68 average 31.5	Anon., FMC Corp. (1973) Caropreso et al. (1974)
St. Augustine, FL full-scale plant	12 (basis unclear)	Anon., FMC Corp. (1976)
San Jose, CA lab-scale, fill-and-draw	40–200	Strunk and Shapiro (1976)
Princeton, NJ small full-scale	100[b]	Caropreso et al. (1974)
Wilmington, DE lab-scale	40–200[b]	Cole et al. (1973)
Washington, DC pilot plant	20–40[b] 200–400[a]	Cole et al. (1973)
Textile mill	60 (basis unclear)	Keller and Cole (1973)

[a]Based on wastewater inflow.
[b]Based on volume of aeration basin and secondary clarifier (single dose).

ative studies of these two chemicals have been conducted, this observation cannot be taken as conclusive.

Keller and Cole (1973) state that the time required to reduce SVI to 50% of its initial value (actual SVI values were not given) was a function of the H_2O_2 dose. Thus, at a dose of 0.4 kg H_2O_2/kg MLSS, day, 50% SVI reduction occurred in less than 1 day, while at a dose of 0.1 kg H_2O_2/kg MLSS, day the 50% SVI reduction took over 8 days. On the basis of these data Keller and Cole (1973) state that the minimum effective H_2O_2 dose for bulking control is approximately 0.1 kg H_2O_2/kg MLSS, day. Experiments confirming this value showed SVI control with a dose of 0.15 kg H_2O_2/kg MLSS, day but no effect on SVI with a dose of 0.035 kg H_2O_2/kg MLSS, day.

Hydrogen peroxide has been successful when dosed continuously and on a batch basis; dose points into the aeration basin, to the RAS line, and to the mixed liquor channel between the aeration basin and the secondary clarifier, have been used with success. Like chlorine, excellent initial mixing of H_2O_2 with the activated sludge is necessary for effective destruction of filamentous organisms. One case history (Anon., FMC Corp., 1976), at St. Augustine, Florida, suggested that a greater contact time between dosed H_2O_2 and activated sludge is required for H_2O_2 than for chlorine. Thus, dosing of 20 mg H_2O_2/L to a degritter located at a point in a mixed liquor channel 5 min flow time ahead of the secondary clarifier resulted in foam on the secondary clarifiers and no control of bulking. When the H_2O_2 dose point was relocated to the overflow of one of the two aeration basins that was 15 min flow time ahead of the secondary clarifier, a dose of 12 mg H_2O_2/L (presumably to only part of the mixed liquor flow) was effective in reducing SVI from 580 mL/g to 178 mL/g in 2 days.

The H_2O_2 generally used for dosing is 50% vol H_2O_2. The recommended feed system is illustrated in Figure 63.

The action of H_2O_2 on filamentous organisms is reported to be one of attacking the sheath, thus destroying the filamentous form (Caropreso et al., 1974). The effect observed, regardless of mechanism, is similar to that when chlorine is used, i.e., the filaments break up and become shorter, and cells within the filaments show signs of lysis.

It is claimed that not only does H_2O_2 kill the filamentous organisms causing bulking, but also, in the oxidation/reduction reactions that accompany this disinfection, oxygen is produced that is available for supplementing DO:

$$2H_2O_2 \rightarrow 2H_2O + O_2$$

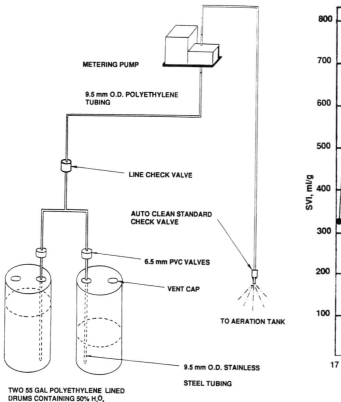

Figure 63. Recommended feed system for dosing hydrogen peroxide. (Anon, FMC Corp., 1976).

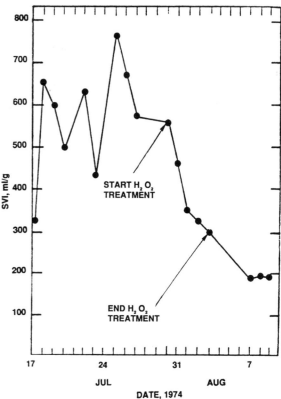

Figure 64. SVI response to hydrogen peroxide treatment at the Petaluma, CA water pollution control plant. (Caropreso et al., 1974).

Should the cause of bulking be due to low DO, then this oxygen should also aid in the specific amelioration of the problem. Conversely, should the activated sludge bacteria develop the ability to rapidly degrade the H_2O_2 prior to its availability for killing filamentous organisms (through production of peroxidase enzymes), its effectiveness for controlling bulking may be compromised.

Case History: City of Petaluma, California

During 1974, the city of Petaluma, California, Water Pollution Control Plant, which treated mixed domestic/industrial wastewater, experienced severe bulking problems with SVI values in the range of 400–700 mL/g (Caropreso et al., 1974). Hydrogen peroxide (as a 50% vol H_2O_2 solution) was dosed to the final quadrant of a two-pass aeration basin that was being fed primary effluent at the 1/4 points and RAS at the head end. Figure 64 shows the doses and concentrations of H_2O_2 used over a 4-day period, and the effect on SVI. The H_2O_2 doses ranged between 9 and 68 mg H_2O_2/L and were variously

applied for 8–24 hr/day. During the incidents, 2300 lb (1050 kg) of H_2O_2 (100% basis) was used to bring the SVI under control (from 550 mL/g to 300 mL/g). The SVI continued to decrease following the termination of H_2O_2 dosing.

SPECIFIC METHODS OF BULKING CONTROL

This section presents specific methods of bulking control rather than the rapid, nonspecific, chemical techniques described previously. The methods described in this section can constitute the second phase of an attack on a bulking problem—the first phase being the use of a rapid, nonspecific chemical method such as chlorination—or they can be used by designing new plants with the features required to implement them. To use the specific methods discussed in this section it is first necessary to characterize the causative filamentous organisms and correlate them with operating and design data to arrive at possible causes for the bulking.

The use of filamentous organism characterization as a diagnostic tool for causes of activated sludge bulking and the association of various filamentous organism types with general and specific sets of conditions in activated sludge has been discussed previously. In this section the relationship between filamentous organism growth conditions (where known) and methods for bulking control will be discussed.

Nutrient Deficiency

In activated sludge, nutrient deficiency almost always means deficiency of nitrogen and/or phosphorus, although some workers have reported deficiencies of "trace" nutrients such as iron (Wood and Tchobanoglous, 1974; Carter and McKinney, 1973). In our experience, nutrient deficiency due to nutrients other than nitrogen and phosphorus is extremely rare. Diagnosis of nutrient deficiency can be by a combination of wastewater analysis and microscopic examination of the activated sludge. If a microscopic examination of the activated sludge reveals the presence of any of the following, then nutrient deficiency is to be suspected:

- Major filament types 021N or *Thiothrix* spp.;
- Viscous activated sludge showing significant amounts of extracellular material by India ink reverse staining or detection of a polysaccharide content of greater than 20% to 25% on a dry weight basis using the anthrone test;
- Foam (scum) on activated sludge aeration basins and secondary clarifiers containing significant amounts of extracellular material (often Neisser positive) and not containing *Nocardia* spp., *M. parvicella* or type 1863.

Poor activated sludge settling due to viscous activated sludge caused by nutrient deficiency cannot be satisfactorily controlled by chlorination or H_2O_2 addition. It can be made to settle with difficulty by large doses of polymer. There is a danger that, should the nutrient deficiency become more severe, the ability to remove soluble organic matter will be lost and the treatment will fail.

In cases where the microscopic examination suggests nutrient deficiency, the BOD_5, COD, or TOC to N and P ratios of the wastewater influent to the aeration basin and the total P and TKN content of the mixed liquor SS should be analyzed. If necessary, the required nutrient(s) should be added. It has often been stated that sufficient nutrients are present when the wastewater to be treated by activated sludge contains $BOD_5/N/P$ in the weight ratio 100/5/1. While these ratios are useful guidelines for detecting nutrient deficiency and for supplementing nutrient-deficient wastes, the following factors also should be taken into account when dealing with nutrient-deficient wastewaters:

- The $BOD_5/N/P$ ratio of 100/5/1 is based on the theoretical maximum N and P needs of bacteria assuming a yield of 0.5g cell biomass per g BOD_5 removed, and a sludge with a 10% N and a 2% P content on a dry weight basis. Nutrient additions less than this ratio are frequently used in some industries. For example, ratios of 100/2.5 to 3.0/0.4 to 0.6 have been used by several papermills with good results.
- The specific nutrient needs of an activated sludge are influenced by sludge age and temperature. Significant nutrient recycling occurs in high MCRT activated sludge due to cell lysis and release of nutrients so that less nutrients per pound of BOD_5 removed are needed than in lower MCRT systems.

  For a given system, nutrient needs per pound of BOD_5 removed increase at lower temperature and decrease at warmer temperature because BOD is used by bacteria for both cell growth and cell maintenance (endogenous respiration). Cell growth on organic carbon (BOD) requires N and P; cell maintenance uses organic carbon without the need for externally supplied nutrients. At warmer temperature, more of the applied BOD is used for cell maintenance energy than at colder temperature, so that the nutrient needs per pound of BOD_5 removed are less at warmer temperature. At colder temperature, more of the BOD is used for cell growth, so that more nutrients are needed per pound of BOD_5 removed. Sludge production (cell yield) also increases at colder temperature for the same reasons. In systems where summer and winter activated sludge temperature varies significantly, a separate summer and winter nutrient addition program should be used.
- The "availability" of N and P in the influent wastewater to the activated sludge aeration basin should be determined. If the waste contains a readily metabolizable carbon

source (e.g., a simple sugar or organic acid) and the major N or P source in the wastewater is organically bound, the nutrients may not be available at a high enough rate for use during the metabolism of the carbon source. An example of such a situation was sometimes encountered at the San Jose/Santa Clara, California, Water Pollution Control Plant in the summer during the years when the plant received a mixture of municipal wastewater and fruit and vegetable canning wastewaters. At times, N supplementation with aqueous ammonia was required. The amount of N supplementation was based on the primary influent TOC/NH_3-N ratio. The primary influent contains organic N as well as NH_3-N but the organic N was discounted when calculating the amount of N supplementation required because it was felt that the organic N may be available at too slow a rate to keep pace with carbon substrate metabolism. Further, because the sewer system is long and the summer sewage temperatures are high, it is likely that some of the organic N in the influent wastewater has already been incorporated into microbial cells by metabolism in the sewer system prior to reaching the plant.

Similar situations have been encountered in some industrial wastewater activated sludge treatment plants. For example, when P bound organically as phytin in a soybean processing wastewater was not available at a high enough rate to supply adequate soluble phosphate, type 021N bulking resulted. The formation of sparingly soluble phosphate precipitates when the aeration basin is operated at pH values near pH = 9 may limit phosphorus availability.

• Nutrient supply should match nutrient demand. If the influent wastewater has wide variations in carbonaceous substrate levels, the nutrient supply must be paced to these variations, at least to the degree that available nutrients never run out in any part of the aeration basin. Often, if this type of situation exists, it is wise to oversupply nutrients during periods of low influent loads so that a "reserve" supply exists in the aeration basin to help meet the demand from shock influent carbon loads. Richard et al. (1985) conducted pure culture chemostat culture work on the nutrient (NH_3-N) limited growth of type 021N and a floc-former isolated from

activated sludge. With a steady-state supply of nutrients, the growth rate of type 021N was never high enough relative to the floc former to become dominant (Figure 65). However, when the NH_3-N supply was intermittent, type 021N showed a much more rapid uptake of NH_3-N than the floc former (Figure 66). This ability allows type 021N to be the successful competitor (and bulking to result) when the NH_3-N supply is intermittent, rather than continuous and matched to the carbonaceous load to a degree that NH_3-N is never depleted. Dual culture chemostat experiments with type 021N under nitrogen limiting conditions (C/N ratio = 15/1) demonstrated that the floc former was selected when the NH_3-N feed was continuous (no bulking), but that type 021N was selected when 75% of the NH_3-N was fed in a spike at 6-hr intervals (bulking) (Figure 67).

• Because each wastewater/activated sludge combination has its own unique nutrient requirements, the suitability of a given influent BOD_5/N/P ratio in meeting these nutrient requirements should be checked by measurements of effluent concentrations of dissolved (0.45 μm filtered) orthophosphate, NH_3-N, and NO_3-N. The required levels will

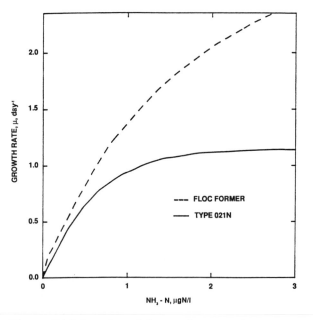

Figure 65. Monod model growth curves for ammonia-limited growth of type 021N and a floc former isolated from activated sludge. (Richard et al., 1985). Reprinted by permission of the Water Environment Federation.

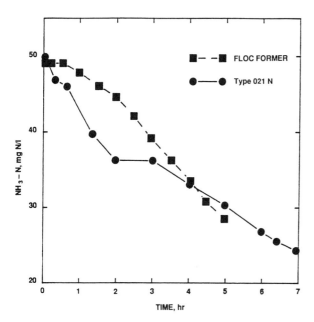

Figure 66. Batch culture ammonia uptake of type 021N and an activated sludge floc former previously grown under ammonia limitation and subjected to a spike in ammonia concentration. (Richard et al., 1985). Reprinted by permission of the Water Environment Federation.

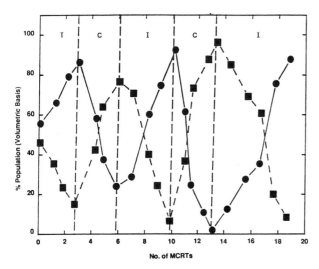

Figure 67. Ammonia-limited growth competition between type 021N (●) and an activated sludge floc former (■) in chemostat culture with intermittent (I) and continuous (C) NH₃-N feeding. C/N = 15; intermittent feeding of 75% of total ammonia at 6-hr intervals. (Richard et al., 1985). Reprinted by permission of the Water Environment Federation.

depend on the needs for maintaining N and P levels in the aeration basin with influent organic load variations. However, concentrations of total soluble inorganic N (NO_3-N + NH_3-N) and soluble orthophosphate, phosphorus of not less than 0.5 to 1.0 mg/L should be maintained. These effluent nutrient levels may not work in all cases, especially when a readily degradable, soluble BOD wastewater is being treated (e.g., papermill, food processing, and brewery wastewaters). Here, a minimum nutrient residual of 1–3 mg/L for both N and P may be required. For example, Reid (1991) and Richard (1991) showed that viscous and filamentous bulking (due to *N. limicola* III) occurred in a Michigan papermill activated sludge system when effluent soluble orthophosphate concentrations were less than 1.0 mg P/L. An effluent soluble orthophosphate concentration of 1–2 mg P/L was needed to avoid these problems.

Both NH_3-N and NO_3-N are available as N sources for activated sludge growth. Therefore, in nitrifying systems, NH_3-N added for nutrient supplementation will be converted to NO_3-N and still remain available as a N source unless anoxic conditions exist somewhere in the system and denitrification of the NO_3-N to N_2 gas occurs.

• Nutrient oversupply can be a problem for N (but not P). Activated sludge plants that dose urea or ammonia and measure only NH_3-N residual in the effluent may end up oversupplying ammonia if nitrification occurs. Because high effluent NO_3-N concentrations (5–10 mg N/L) may result and cause denitrification and sludge flotation in the final clarifier, it is best to regulate addition of N to the activated sludge to maintain a total inorganic N residual (NH_3-N + NO_3-N) at a desired concentration.

Some industrial wastewater activated sludge plants use a commercial agricultural N fertilizer that contains a mixture of urea, ammonia and nitrate (usually 24% to 25% nitrate). Use of this mixture causes problems because the NH_3-N residual is usually nonexistent while the NO_3-N residual builds up to rather high amounts, leading to denitrification. For this reason, formulations without nitrate are recommended.

• In some activated sludge effluents, especially those from pulp and paper waste treatment, the colorimetric analysis of orthophosphate employing the production of a molybdenum blue [e.g., stannous chloride and ascorbic acid reduction methods (*Standard Methods*, 1991)] suffers from a positive interference. This can have serious results if not detected because it can result in the influent phosphate feed being reduced on the mistaken belief that there is residual soluble P in the effluent—this has on one occasion resulted in P deficient bulking that was detected only when it was realized that no phosphate feed was being added, yet soluble ortho-P was still being detected in the effluent. An effluent sample spiked with ortho-P should always be run to check for P recovery.

Brown-colored lignins in papermill systems frequently interfere in the colorimetric tests for nitrate. Here, a NO_3^- ion-specific probe is recommended for monitoring effluent NO_3-N concentration.

Low Dissolved Oxygen (DO)

Low DO concentrations can cause the growth of several filamentous organism types in activated sludge. At low to moderate MCRT values, *S. natans* and type 1701 certainly are associated with low-DO bulking, and it is possible that *H.*

hydrossis also may cause bulking under these conditions. Studies by Hao (1982) and Richard et al. (1985) suggest that type 1701 may occur at more severe DO limitation than *S. natans*. At long MCRT values, *M. parvicella* growth is favored by low DO (Slijkhuis, 1983).

Palm et al. (1980) conducted laboratory experiments on low DO bulking using settled domestic sewage feed and continuously-fed completely-mixed aeration basins. They found that the DO concentration required to prevent the growth of *S. natans* (the filamentous organism that caused bulking in their experiments) was a function of the F/M—the higher the F/M, the greater was the DO concentration required. Their experiments led to a relationship between the F/M and aeration basin DO concentration to prevent low DO bulking for a completely mixed aeration basin (Figure 68). In Figure 68, areas of F/M and aeration basin DO concentration are shown where bulking did and did not occur. In addition, a safe operating line is shown (plotted at +0.5 mg DO/L from the center point DO concentration of the region of nonbulking). This line is plotted because in some of the experiments the SVI of the "nonbulking" sludge tended to increase to levels that would be unacceptable in practice.

Palm et al. (1980) showed that low-DO bulking

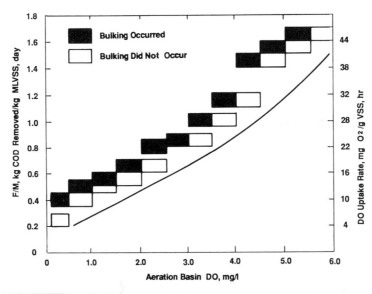

Figure 68. Combinations of F/M and aeration basin DO concentration where bulking and non-bulking sludge occurs in completely-mixed, continuously-fed aeration basins. (Palm et al., 1980). Reprinted by permission of the Water Environment Federation.

could be caused and cured by manipulation of F/M and DO concentration (Figure 69). The onset of bulking tended to be much more rapid than its cure. In fact, it was demonstrated that curing low-DO bulking by increasing aeration basin DO concentration was achieved by washout of the filamentous organism population, and required a period of 3 MCRTs. Thus, for a plant with a 10-day MCRT, 30 days would be required to reach the new and lower steady-state level of filamentous organisms associated with the increased aeration basin DO level. This finding emphasizes the importance of having a rapid method for decreasing filamentous organism population (such as chlorination).

The safe operating line presented in Figure 68 is specific for the experiments conducted by Palm et al. (1980), and the results should only be applied in principle rather than in detail to other systems. Different wastewater and activated sludge combinations will typically have different DO uptake rate-F/M relationships. This is important because Lau et al. (1984a and 1984b) showed that the DO uptake rate was an important factor in determining the competitive growth of *S. natans* and an activated sludge floc-forming

organism. Palm et al. (1980) developed a relationship between the DO uptake rate, mgO$_2$/g SS, hr, and F/M (Figure 70) that can be used in generalizing the data to systems with different DO uptake rate-F/M relationships. The data presented in Figure 70 relating F/M and DO uptake rate have been used to construct the right-hand scale in Figure 68 so that the DO concentration required to suppress low DO bulking in a completely mixed activated sludge system can be read as a function of DO uptake rate.

The utility of this relationship is shown by experience at the Orange County Sanitation District's Oxygen Activated Sludge Plant at Fountain Valley, California. At the time the following data were collected, this plant treated a mixed domestic and industrial wastewater stream which on occasion contained citrus processing wastewater. Oxygen uptake rates in the first aeration basin compartment reached values of approximately 200 mg O$_2$/g VSS,hr. When the plant was operated at DO levels of 10–12 mg/L in the aeration basin, bulking due to type 1701 occurred. Note that this organism is associated with "low DO" bulking, even though the DO was 10–12 mg/L. This is possible because the very high DO uptake

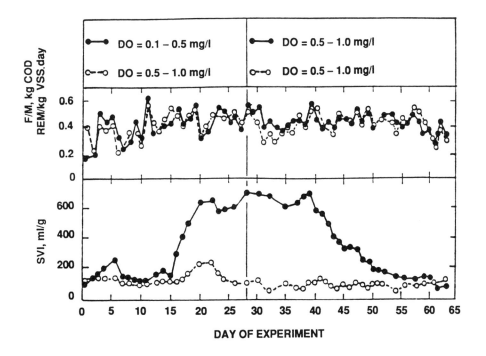

Figure 69. Temporal variation of F/M and SVI at aeration basin DO values of 0.1 to 0.5 mg/L and 0.5 to 1.0 mg/L. (Palm et al., 1980). Reprinted by permission of the Water Environment Federation.

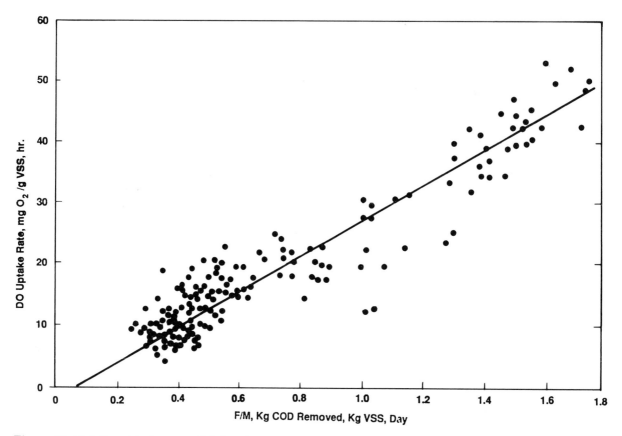

Figure 70. Relationship between F/M and DO uptake rate of activated sludge from a completely mixed system. (Palm et al., 1980). Reprinted by permission of the Water Environment Federation.

rates required very high bulk DO concentrations to prevent the growth of low DO filaments. When the aeration basin DO concentration was raised to 16–20 mg/L, type 1701 levels decreased and sludge settling improved.

In another example of this type of relationship, the SVI of a pulp and paper waste treatment activated sludge was found to be related to the DO uptake rate of the activated sludge (Figure 71). The filamentous organisms present and causing the settling problems were largely the low DO types: *H. hydrossis* and type 1701. Figure 71 shows that the SVI increased approximately linearly with DO uptake rate. At this plant it had been previously observed that SVI increased with increased mixed liquor temperature. Since mixed liquor DO uptake rates increase with increased mixed liquor temperature, the observation that high temperatures caused poor settling sludge could also be explained by the occurrence of low DO bulking.

In dealing with low-DO bulking problems, a

basic economic question must be addressed. To cure the problem specifically, one must either lower the F/M or increase the aeration basin DO concentration. Each of these steps may have undesirable consequences. Lowering the F/M may:

• Cause onset of nitrification, which will impose additional oxygen demands
• Increase carbonaceous oxygen requirements due to increased endogenous respiration
• Increase MLSS levels to values that exceed the solids handling capacity of the secondary clarifiers.

Increasing the aeration basin DO concentration certainly will require greater power input (energy consumption) and may cause the onset of nitrification in systems with high enough MCRT values.

Because of these factors it is necessary to consider the alternatives of:

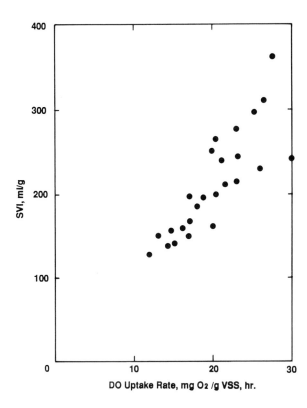

Figure 71. Effect of DO uptake rate on SVI of a pulp and paper waste activated sludge treatment plant (monthly average data).

1. Continuing to operate at low aeration basin DO concentration and killing off the filamentous organisms that develop using RAS chlorination
2. Considering modification of the aeration basin from a completely mixed to a "selector" configuration (since selector systems are effective in controlling all low DO filamentous organisms with the exception of *M. parvicella*).

The use of selectors will be discussed later. The former approach was used effectively by Banoub (1982) at the Woonsocket, Rhode Island activated sludge plant to provide aeration power savings. At the time of Banoub's work, the plant flow was 8.5 MGD (0.37 m³/sec), while the design flow was 16 MGD (0.7 m³/sec). Because of the high cost of electrical energy ($0.08/kwh, including fuel adjustment), it was decided to economize on aeration and maintain a low aeration basin DO. This encouraged the growth of low DO filamentous organisms and an accompanying increase in SVI. A target SVI of 225 mL/g

was established, and RAS chlorination at a rate 3 kgCl₂/10³ kg MLSS,day was used to reduce the SVI to below this value on the few occasions that it was exceeded. Chlorination took "a couple of days" to reduce the SVI to the target value; exposure frequency of the solids inventory to chlorine was 8 times per day.

Previous operation to maintain an aeration basin DO of 2.0 mg/L had required the operation of one aeration blower at maximum amperage (330 amps) during the winter, and an additional aeration blower during the summer. When operating at low DO (0.2 to 0.5 mg/L) only one aeration blower operating at its minimum amperage (240 amps) was required. This reduction in aeration requirements resulted in a yearly savings of over $41,000 (average of $112/day). On the days when RAS chlorination was necessary, approximately 85 lb Cl₂/day (39 kg/day) was used, at a cost of less than $10/day.

Banoub (1982) pointed out that extreme care must be taken when operating in such a mode. Frequent monitoring of activated sludge settleability and microscopic examinations of activated sludge were made. Banoub also indicated the exceptionally high effluent quality (on the order of 5 mg/L each of BOD₅ and SS, and turbidity of 2 FTU) attainable from a high SVI sludge. It should be remembered, however, that the plant was receiving approximately 50% of its design flow, so that ample secondary clarification capacity was available to accommodate the high SVI sludge.

Aeration Basin Configuration and Method of Wastewater Influent Feeding

In general, activated sludge systems that are continuously fed and have well-mixed aeration basins that are aerated throughout will produce poorer settling activated sludge than systems that are fed intermittently, or have compartmentalized aeration basins where there is a relatively high initial concentration of wastewater at the point where the RAS and the influent wastewater mix (Heide and Pasveer, 1974; Rensink et al., 1982; Tomlinson, 1982; and Chiesa and Irvine, 1985). Furthermore, activated sludge settling usually is poorer if this initial mixing or contact zone between RAS and influent wastewater contains DO and/or is aerated rather than being devoid of DO and/or unaerated (Chiesa and Irvine, 1985).

Tomlinson and Chambers (1978b) conducted a survey of full-scale activated sludge plants in Great Britain in which SSVI₃.₅ (the stirred SVI at

SS = 3.5 g/L) and the distribution of hydraulic residence times in the aeration basin alone (without secondary clarifier and RAS stream) were measured. The tracer study results for hydraulic residence time distribution were interpreted using a completely-mixed tanks in series model (i.e., one tank = completely-mixed; infinite number of tanks = plug flow). It was found that as the degree of "plug flowness" increased, the $SSVI_{3.5}$ generally tended to lower, less erratic values (Figure 72). Heide and Pasveer (1974), working with low loaded, completely nitrifying oxidation ditches and a 24-hr fill-and-draw laboratory batch activated sludge unit, showed that the SVI was from 500 to 600 mL/g for a continuously-fed oxidation ditch, 100 to 150 mL/g for an oxidation ditch fed discontinuously, and from 40 to 50 mL/g for the laboratory batch unit fed once per day. When an oxidation ditch was preceded by a small anoxic mixing tank (hydraulic detention time of 15 min) through which both the RAS and wastewater passed, the SVI was less than 70 mL/g.

The occurrence of bulking in completely-mixed systems but not in "plug-flow" or intermittently-fed systems has been variously explained. Chudoba et al. (1973a) proposed that filamentous organisms possess lower half-saturation coefficients (K_s) for organic carbon substrates than floc-forming microorganisms. This allows their selection in completely-mixed, low F/M systems where the carbonaceous substrate concentration is low throughout the aeration basin because the influent wastewater is dispersed throughout its entire volume. This hypothesis cannot completely account for the suppression of bulking that is achieved by allowing the activated sludge to pass through a zone of high carbonaceous substrate concentration for a very short time period prior to entering the aeration basin. It has been proposed that the effectiveness of such systems in suppressing bulking is because they favor the growth of floc-forming organisms with higher substrate uptake rates and storage capacities (a transient, unbalanced growth response) than filamentous organisms (van den Eynde et al., 1982b; Grau et al., 1982), or of floc formers that have greater starvation resistance than filamentous organisms (Chiesa and Irvine, 1985).

Control of bulking has been obtained consistently in laboratory studies by introducing plug flow characteristics into the aeration basin (Rensink et al., 1982; Eikelboom, 1982a), intermittent feeding of wastewater (Houtmeyers, 1978; Houtmeyers et al., 1980; Verachtert et al., 1980; van den Eynde et al., 1982a; 1982b), compartmentalization of aeration basins (Chudoba et al., 1973a, 1973b; 1974; Spector, 1977), use of a small mixing tank(s) where RAS and influent wastewater mixes prior to the aeration basin (Lee et al., 1982; Grau et al., 1982), or fed-batch operation (Goronszy, 1979; Goronszy and Barnes, 1979; Barnes and Goronszy, 1980; Chiesa and Irvine, 1985). All of these measures produce a carbonaceous substrate concentration gradient through the aeration basin, or a high substrate concentration at the point where RAS and influent wastewater enter the aeration basin system, followed by close to zero soluble substrate concentration thereafter in the aeration basin.

Chudoba et al. (1973) first coined the term "selector" to describe the activated sludge system containing a small separate initial mixing zone(s) for RAS and influent wastewater (Figure 73). The term "selector" refers to the role of such a device in "selecting" activated sludge organisms with desirable settling characteristics. The need for compartmentalization (a cascade) of these initial contact zones was stressed.

The beneficial effect of anaerobic or anoxic conditions in the initial contact zones was recognized in some of the early work on the effect of aeration basin configuration on activated sludge settling. For the purposes of this discussion the following definitions will be used:

Aerobic or Oxic: DO present and supplied in sufficient quantities to demands established by

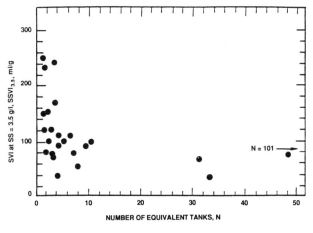

Figure 72. Relationship between stirred sludge volume index at SS = 3.5 g/L and aeration basin mixing characteristics (Tomlinson and Chambers, 1978b).

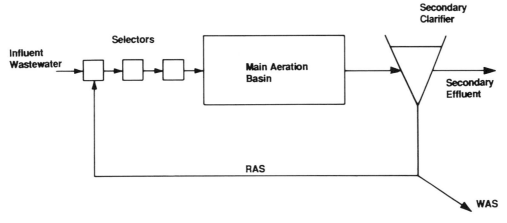

Figure 73. "Selector" system configuration.

metabolic activity. Aerobic respiration and storage are primary mechanisms functioning for soluble BOD removal.

Anoxic: DO absent; NO_3-N present and supplied in sufficient quantities to meet demands established by metabolic activity. Denitrification and storage are primary mechanisms functioning for soluble BOD removal.

Anaerobic: DO and NO_3-N absent. Stored inorganic polyphosphate depolymerization (hydrolysis) is the major metabolic activity removing soluble BOD.

Enhanced biological phosphorus removal (EBPR) by activated sludge employs a sequence of an initial anaerobic zone(s) followed by an aerated or oxic zone(s). One of the first patents for EBPR was in fact titled "Production of Non-Bulking Activated Sludge" (U.S. Patent 4,056,465; Spector, 1977). This patent provides the basis for the so-called "A/O" process. Plants employing the A/O process usually have activated sludge with excellent settling properties (Hong et al., 1984; Deakyne et al., 1983). Tracy et al. (1986) demonstrated that anaerobic conditions provided an additional improvement of settling properties over that afforded by aeration basin compartmentalization. These workers ran fill-and-draw activated sludge systems in which the feeding conditions were anaerobic, aerobic, and oxygen-limited. These feed periods were all followed by an oxic period, settling, and effluent withdrawal. With initial anaerobic contact the SVI was 55–65 mL/g; with the initial oxic contact period the SVI was 100 mL/g; when the initial

contact period was oxygen-limited the SVI was 300 mL/g, most likely due to low DO filamentous bulking. Similar results were obtained by Watanabe et al., 1984; Inamori et al., 1986; and Chiesa and Irvine, 1985.

Initial anoxic zones were found to control bulking when NO_3-N was generated in the aerated part of the aeration basin (Cooper et al., 1977). Tomlinson and Chambers (1979) concluded that an initial anoxic zone placed in a nitrifying activated sludge system improved settling properties both by being anoxic and by reducing longitudinal mixing. Wagner (1982) improved activated sludge settling in nitrifying activated sludge by introducing a predenitrification reactor. Hoffman (1987) showed that compartmentalized aeration basins with an initial anoxic zone showed better settling activated sludge than an equally compartmentalized, totally oxic system. Chiesa and Irvine (1985) and Ip et al. (1987) demonstrated filamentous organism suppression in sequencing batch reactors with an anoxic fill period.

The Selector Effect

When activated sludge is subjected to a sequence of environmental conditions where it is first fed and then starved as in a compartmentalized reactor, a batch fill-and-draw system or a selector system (Figure 73), the microbial make-up of the sludge adjusts to this situation by developing the ability to rapidly take up soluble substrate and store it internally for use later during the "starve" period. This soluble substrate uptake and storage ability is the basis of the "selector effect" because many filamentous organisms are less able to perform in this way than some members of the floc-forming community. The selector effect is illustrated in Figures 74 and 75. These curves compare the time course of soluble substrate concentration (soluble COD) and respiration rate (DO uptake rate) in a batch test on activated sludge taken from a selector activated sludge system and from a completely mixed (CSTR) acti-

vated sludge system (van Niekerk et al., 1987). The selector activated sludge takes up soluble COD much faster than the CSTR activated sludge (Figure 74). The DO uptake rate increases while the rapid soluble COD uptake is occurring in the selector activated sludge (Figure 75). Measurements of the total mass of soluble COD taken up and of the total mass of DO consumed by the

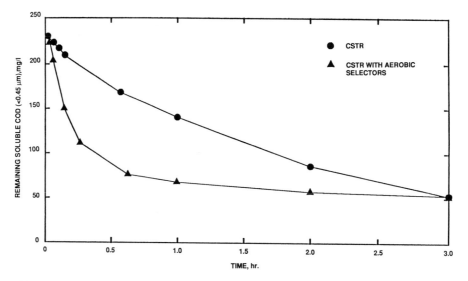

Figure 74. Soluble COD uptake by activated sludge during batch substrate uptake experiments on activated sludge from CSTR and aerobic selector systems. (van Niekerk et al., 1987). Reprinted by permission of the Water Environment Federation.

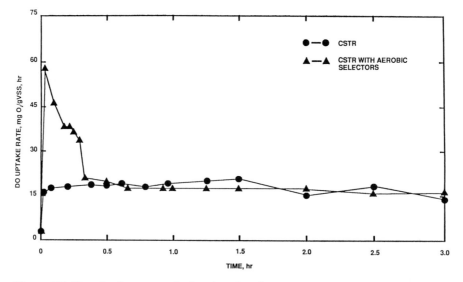

Figure 75. Respiration rates during batch substrate uptake experiments on activated sludges from CSTR and aerobic selector systems. (van Niekerk et al., 1987). Reprinted by permission of the Water Environment Federation.

selector activated sludge during this initial period would show that only a small fraction of the soluble COD that disappears from solution (usually less than 25% and often as little as 10%) can be accounted for by oxygen consumption. This means that the majority of this soluble COD is not oxidized — rather, it is stored.

It is important to realize that selector systems not only select *against* some filamentous organisms, but they also select *for* certain types of organisms (happily, mostly nonfilamentous) that are able to take up soluble substrate rapidly and store it. Therefore, to survive in a selector system (and therefore become selected!!) an organism must have a high soluble substrate uptake rate and a large substrate storage capacity. These characteristics are not found in the floc-forming microorganisms that grow in a completely-mixed, continuously-fed activated sludge system (Figure 74). Therefore, it must be assumed that selectors change the type of floc-forming organisms as well as deselecting against the filamentous organisms. Evidence that this is so has been provided by van Niekerk et al. (1987) who showed that, in aerobic selectors, large numbers of amorphous zoogloeal colonies (Figure 12d) developed. These colonies were not seen in CSTR activated sludge; the colonies were isolated and shown to be capable of high soluble substrate uptake rates and to have large substrate storage capacities.

Figure 76 illustrates the ability of two types of organisms isolated from activated sludge to take up acetate (a simple, soluble substrate) in batch culture — conditions that simulate a selector reactor (van Niekerk et al., 1987). *Zoogloea ramigera* is a floc-forming bacterium typically found in aerobic selector activated sludge. Type 021N is a filamentous organism that grows well on acetate and can cause bulking in activated sludge treating wastewater with high soluble COD concentrations. The acetate uptake rate of *Z. ramigera* is higher than that of type 021N over most of the growth rate range tested. Figure 76 predicts that under intermittent feeding conditions (typical of a selector), *Z. ramigera* almost always would be able to sequester the acetate before type 021N had a chance to get at it. The picture is quite different when acetate is fed continuously (typical of a steady-state CSTR-activated sludge system). Figure 77 shows that at low acetate concentrations, type 021N can outcompete *Z. ramigera*. van Niekerk et al. (1987) showed that these competitive advantages could actually determine which of these organisms dominated in a mixed chemostat culture grown at a steady state growth

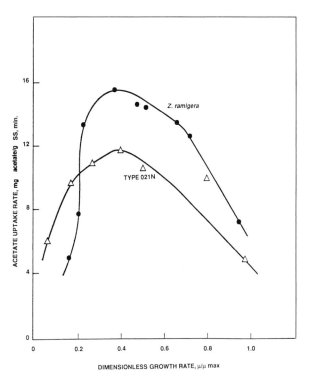

Figure 76. Maximum acetate uptake rates of *Z. ramigera* and type 021N in batch tests. (van Niekerk et al., 1987). Reprinted with permission of the Water Environment Federation.

rate designed to favor the filamentous organism, type 021N. When a low frequency, intermittent feeding of acetate was employed, *Z. ramigera* dominated (nonbulking); for continuous feeding and high frequency intermittent feeding, type 021N dominated (bulking) (Figure 78).

Inherent in the above description of selector mechanisms is the fact that selectors function only on soluble substrate. It is probable that we can be even more restrictive by stating that selectors only influence that part of the soluble substrate that can be readily transported across the microbial cell wall and readily converted into intracellular storage products. Very likely this confines selector function to removing the very simple, readily-metabolizable organic substances. This point is illustrated in Figures 79 and 80 (Shao and Jenkins, 1989) that show the relationship between the soluble substrate concentration leaving a selector and the SVI of the activated sludge. When soluble substrate is measured as soluble COD (0.45 μm filtered COD), a selector effluent concentration of up to about 60 mg/L will allow an SVI of ≤ 100 mL/g to be achieved (Figure 79). However, if the simple, readily trans-

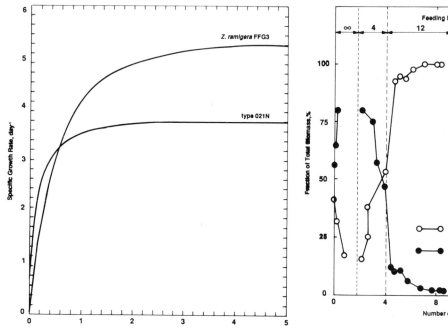

Figure 77. Predicted curves relating steady state growth rate to acetate concentration for type 021N and *Z. ramigera*. (van Niekerk, et al., 1988). Reprinted by permission of the Water Environment Federation.

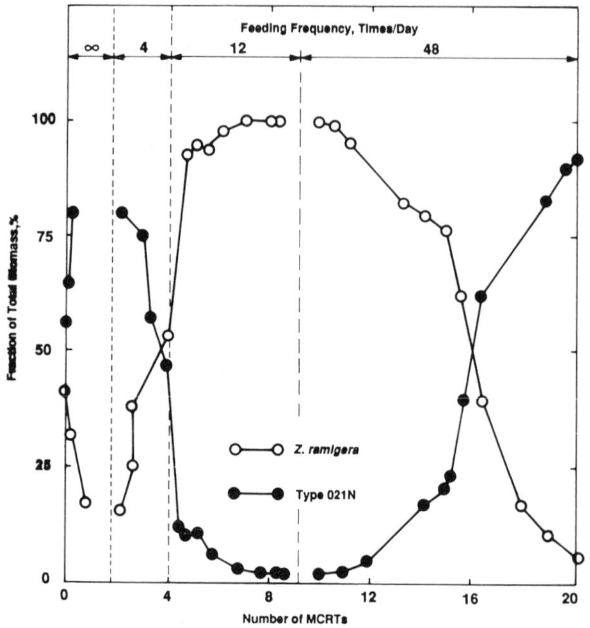

Figure 78. Dual-culture chemostat growth of *Z. ramigera* and type 021N with intermittent feeding of carbon substrate (van Niekerk, et al., 1987). Reprinted by permission of the Water Environment Federation.

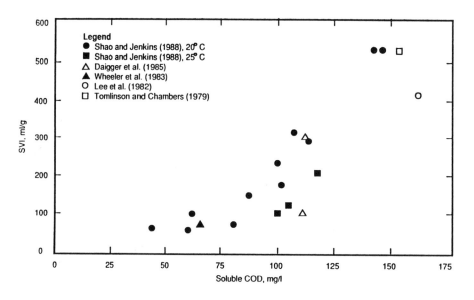

Figure 79. SVI as a function of selector effluent soluble COD concentration. Reprinted with permission from Shao and Jenkins, "The use of anoxic selectors for the control of low F/M activated sludge bulking," *Water Sci. Technol.*, 21, 1989. Pergamon Press plc.

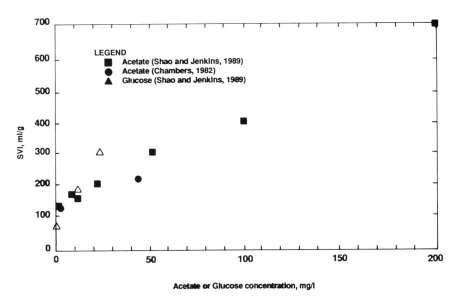

Figure 80. SVI as a function of acetate or glucose concentration in anoxic selector effluent. Reprinted with permission from Shao and Jenkins, "The use of anoxic selectors for the control of low F/M activated sludge bulking," *Water Sci. Technol.*, 21, 1989. Pergamon Press plc.

portable and metabolizable organic components of the soluble COD are measured in the selector effluent, one sees that these must be almost completely removed (<1 mg/L) to achieve an SVI of 100 mL/g (Figure 80).

When organic substrates are transported into microbial cells and transformed into storage products, a source of energy is required. This energy is generated by oxidizing part of the substrate. As indicated previously, selectors are classified into aerobic, anoxic, and anaerobic on the basis of the method of achieving this oxidation. If DO is used, the selector is aerobic (Figure 81); if NO_3-N is used, the selector is anoxic and denitrification occurs (Figure 82); if hydrolysis of internally-stored high energy polyphosphate is used, the selector is anaerobic and release of orthophosphate occurs. In the latter case, EBPR occurs when released and influent orthophosphate is subsequently taken up in the downstream main aeration basin using energy generated through the oxidation of stored substrate (Figure 83).

The selector substrate uptake mechanisms discussed above are virtually mutually exclusive. This happens because of the competitive environment of activated sludge. The energy yield obtained by the oxidation of a unit amount of an organic substrate by the processes of aerobic oxi-

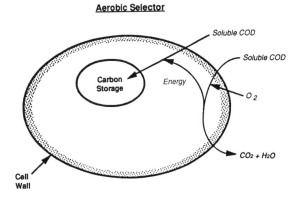

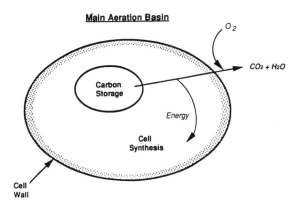

Figure 81. Aerobic selector substrate uptake mechanism.

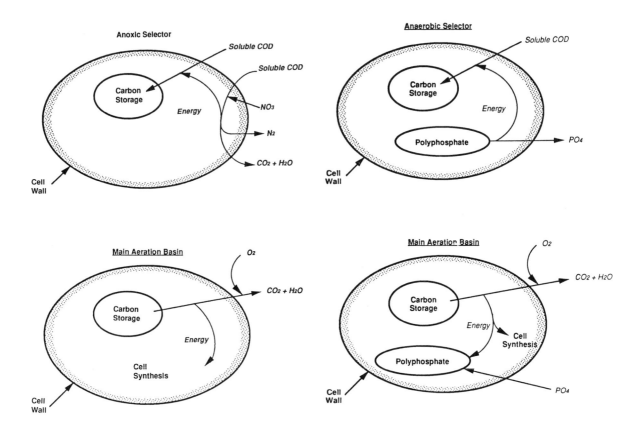

Figure 82. Anoxic selector substrate uptake mechanism.

Figure 83. Anaerobic selector substrate uptake mechanism.

dation (where O_2 is the terminal electron acceptor) is greater than for denitrification (where NO_3 is the terminal electron acceptor), and in turn this is greater than for polyphosphate hydrolysis (the electrons are "stored" internally as polyphosphate to be later accepted by O_2 in the aerobic aeration basin). Thus, if there is DO present, a mixed microbial community such as activated sludge will become dominated by microorganisms that perform aerobic oxidation. Denitrification will not occur, even if NO_3 is present, because it is less efficient than aerobic oxidation. Only when there is no DO available will denitrification occur. Likewise, the presence of both DO and NO_3 will suppress polyphosphate storage and hydrolysis. The members of the microbial community only carry out the lower efficiency oxidations when they are forced to; i.e., when it is the only game in town. In many instances, however, more than one type of metabolism can occur in different portions of the same basin. For example, both aerobic and anoxic metabolism can occur in an initial mixing cell which is not uniformly aerated.

Likewise, anoxic metabolism may occur in the upstream portion of a mixed unaerated initial contact basin, but anaerobic metabolism will occur in the downstream portion if the supplied NO_3-N is exhausted, and readily degradable soluble substrate remains. The activated sludge floc itself can also provide for an environment where more than one type of metabolism can occur concurrently. For example, in a highly-loaded, poorly-aerated initial contact zone, aerobic oxidation can occur in the bulk liquid and at the floc surface, while denitrification and possibly anaerobic phosphate release take place inside the activated sludge floc.

Another aspect of the type of energy metabolism occurring in the selector is its effect on the selection process itself. Most activated sludge filamentous organisms are aerobes. Because of this, in an aerobic selector the selection process occurs based only on the relative ability of the microorganisms to rapidly take up and store substrate. This type of selection is called "kinetic." When either anoxic or anaerobic conditions are imposed on the selector, the kinetic selection is supplemented by "metabolic" selection — either the ability to denitrify or the ability to store and hydrolyze intracellular inorganic polyphosphate. The denitrification rate of some filamentous organisms is much lower than the floc-forming bacteria growing in selectors (*Z. ramigera*) (Table 28) (Shao and Jenkins, 1989; Blackall et al., 1991). Moreover, some of these filamentous organisms (e.g., type 021N and *Nocardia amarae*) only reduce nitrate to nitrite rather than all the way to nitrogen gas, as *Z. ramigera* does (Blackall et al., 1991).

Some types of floc-forming bacteria (e.g., *Acinetobacter* spp.) can take up and store substrate using polyphosphate hydrolysis for energy generation. It is suspected that many types of filamentous organisms are unable to do this. Blackall et al. (1991) have shown *Nocardia amarae* is unable to take up substrate anaerobically.

In discussing the selector effect thus far we have only made reference to the events occurring in the initial feeding zone. For proper functioning, events that occur in the aeration basin also are very important. The fact that storage takes place in the selector can be demonstrated in several ways. Shao and Jenkins (1989) measured the

decay of various types of activated sludge activity for sludges developed in anoxic selectors and CSTR systems when they were aerated in the absence of exogenous substrate (as they would be in the aeration basin following a selector). As Table 29 shows, the anoxic selector sludge activities persisted longer than in the CSTR sludges. Combined with the observation that the MLSS decay rate for anoxic selector sludge was lower than for CSTR sludge, this suggests the presence of storage products that allowed the selector sludge floc formers to survive for longer periods of time in the absence of exogenous substrate than the CSTR floc formers.

Besides being able to better survive in the aeration basin under "starvation" conditions, the organisms that took up and stored substrate in the selector must be able to utilize their stored products. If they do not, they will return to the selector with their storage capacity diminished. This situation could eventually lead to a saturation of storage capacity and leakage of soluble organic substrate through into the aeration basin. If this occurred, the feed/starve cycle required for the selector effect would be eliminated, and filamentous organism growth could be expected. It is, therefore, important to not

Table 28. Denitrification Rates for Type 021N, *N. amarae* Strains, *Z. ramigera* (Blackall et al., 1991) and an *M. parvicella*-Dominated Activated Sludge (Wanner and Grau, 1988)

Strain	"Type" of Organism	Denitrification Rate, mg NO_3-N/gSS, min
type 021N	Filament (bulking)	8.3×10^{-4}
N. amarae ASF3	Filament (foam)	3.2×10^{-4}
N. amarae ASAC1	Filament (foam)	2.4×10^{-3}
Z. ramigera	Selector floc former	3.3×10^{-1}
M. parvicella-dominated activated sludge	Filament (bulking and foaming)	$1.8-2.7 \times 10^{-1}$

Table 29. Comparison of Decay Rates for Activated Sludges Under Starvation Conditions from Completely Mixed and Anoxic Selector Systems (Shao and Jenkins, 1989)

Activated Sludge Type	"First order" decay rate, k_d(day^{-1}) based on:			
	MLSS	COD Uptake Rate	Oxygen Uptake Rate	NO_3-N Uptake Rate
CSTR System	0.050	0.39	0.38	0.57
Anoxic Selector System	0.034	0.25	0.22	0.29

Table 30. Effectiveness of Selectors in Controlling Filamentous Organisms

Selectors Effective	Selectors Not Always Effective
S. natans	type 0041
type 1701	type 0675
type 021N[a]	type 0092
Thiothrix spp[a].	*M. parvicella*
N. limicola	
H. hydrossis	
type 1851	
Nocardia spp.[b]	

[a]Not when caused by nutrient deficiency.

[b]Aerobic selectors not always effective; anoxic selectors effective (see Cha et al., 1992).

only size a selector correctly to encourage the rapid uptake of the dissolved organic compounds, but also to design the aeration basin so that it can provide proper conditions for utilizing the stored substrate present in the microorganisms.

From the previous discussion it can be deduced that selectors should be generally effective against filamentous organisms that grow on simple soluble organic compounds (such as are removed in selectors), against filamentous organisms that do not have high soluble substrate uptake rates, and against filamentous organisms that do not have high substrate storage capacities. In addition to this, filamentous organisms that do not denitrify rapidly or completely (or at all), and filamentous organisms that are not able to take up substrate coupled with polyphosphate hydrolysis will be outcompeted in anoxic and anaerobic selectors, respectively. Selectors are not effective against the growth of filamentous organisms caused by low pH, septic or other sulfide-containing wastewaters, and nutrient (N and P) deficiency.

Table 30 lists the types of filamentous organisms against which selectors have been successful. Also included in Table 30 are the filamentous organisms against which selectors are not consistently effective. The reasons for the inability of selectors to control these filamentous organisms is not well understood. Some hypotheses are:

1. The filamentous organisms have high substrate uptake and storage capacity (possibly *M. parvicella*)
2. The filamentous organisms grow on either complex organic matter or particulate material whose concentration is not influ-

enced in a selector system (possibly type 0675, type 0041, and type 0092).
3. The filamentous organisms denitrify (possibly *M. parvicella* and type 0092). Table 28 shows that the denitrification rate of an activated sludge dominated by *M. parvicella* is of the same order of magnitude as the selector floc-former *Z. ramigera*.

The filamentous organisms listed above all grow in long MCRT (>15 days) activated sludge plants, and especially in plants that incorporate unaerated (anoxic and anaerobic) zones—typically biological nitrification/denitrification and EBPR plants.

Finding out the causes and controls for these filamentous organisms is an important area for research. Biological N and P removal plants are becoming widespread in the USA. Most of these filamentous organisms grow inside the flocs producing a diffuse floc structure; for this situation, control by RAS chlorination can require high and prolonged chlorine doses which can cause some deterioration in effluent quality and can, more importantly, compromise EBPR.

Selector Design (Sizing)

Early approaches to selector design involved using selectors sized on the basis of a certain fractional volume of the aeration basin. This was derived from the work of investigators such as Lee et al. (1982), who showed that activated sludge settling characteristics were a function of fractional selector size (Figure 84). This approach has been refined considerably in recent years. The key points that must be addressed for a successful selector design are:

1. Establish a high enough initial contact zone

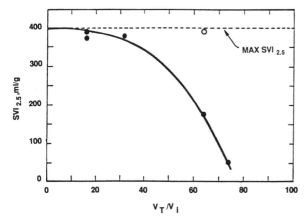

Figure 84. Relationship of $SVI_{2.5}$ to relative size of initial aeration basin compartment. V_t = total aeration basin volume; V_i = initial aeration basin volume. Reprinted with permission from Lee et al., "The effect of aeration basin configuration on activated sludge bulking at low organic loading," *Water Sci. Technol.*, 14, 1982. Pergamon Press plc.

organic loading [$(F/M)_i$] to produce rapid soluble organic matter uptake rates. Especially useful in this regard is the work of Tomlinson (1976) showing the relationship between SVI and initial contact zone organic loading (Figure 85). These data

suggest that $(F/M)_i$ values of greater than about 3 kg BOD_5/kg MLVSS,day will produce an SVI value in the neighborhood of 100–150 mL/g. This is critical for aerobic selectors, but less important for anoxic and anaerobic selectors.

To provide a margin of safety, the first compartment of an aerobic selector should be designed for an organic loading of approximately 6 kg BOD_5/kg MLSS,day (or approximately 12 kg COD/kg MLSS,day).

2. A selector should be sized to allow sufficient time for the removal of all of the soluble, readily transportable, and metabolizable organic matter. Since it usually is not possible to measure the exact amount of this type of organic matter, soluble COD is used as a surrogate parameter. The data of Shao and Jenkins (1989) suggest that (for a domestic wastewater) a selector effluent soluble COD of approximately 60 mg/L will produce an SVI of about 100 mL/g (Figure 79). Selector design practice in Czechoslovakia and Austria (Chudoba and Wanner, 1987) uses the COD removal criterion of "80% of the removable COD" in the selector where removable COD is defined as influent soluble COD minus secondary effluent soluble COD.

3. Because of the need to reduce longitudinal mixing, and because waste flow and strength often vary widely from average values, it is good practice to provide several compartments in a

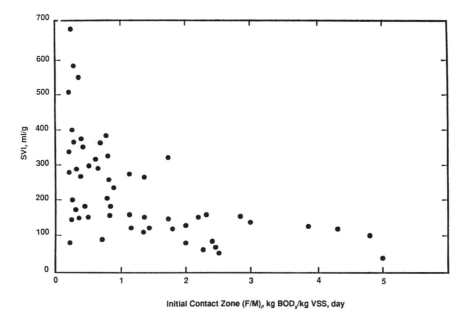

Initial Contact Zone (F/M)ᵢ, kg BOD₅/kg VSS, day

Figure 85. Effect of initial contact zone F/M on SVI for aerated initial contact zones (Tomlinson, 1976).

selector. It is usual to provide a minimum of 3 compartments. For aerobic selectors these compartments can be sized on the basis of the COD load to the first selector compartment as follows:

First compartment   12 kg COD/kg MLSS,day
Second compartment   6 kg COD/kg MLSS,day
Third compartment   3 kg COD/kg MLSS,day

This procedure yields a 3-compartment selector with the first two compartments equal in size and the third compartment double the size of the first compartment.

Initial contact zone organic loads much in excess of these values should be avoided (especially with wastes containing significant amounts of readily available organic matter), because viscous activated sludge may be produced. In the first compartment of an aerobic selector in activated sludge plants treating brewery wastewater and soft drink bottling wastewater, DO uptake rates in excess of 200 mg O_2/g VSS,hr were produced. Viscous (slime) bulking resulted. Bypassing the selector eliminated the viscous bulking.

4. For aerobic selectors, a supply of oxygen is required. Oxygen uptake rates of 50–60 mg O_2/g MLVSS,hr or more can be encountered; however, the amount of oxygen required is only a small fraction of the soluble COD removed—typically only 15% to 25%. The DO concentration should be between 1 to 2 mg/L.

5. For anoxic selectors, the $(F/M)_i$ values can be reduced because the selector utilizes both a feed-starve cycle (kinetic) as well as the ability to denitrify (metabolic) for selection against filamentous organisms. For the same reason, compartmentalization of the selector to produce high initial F/M ratios is also less important. In fact, experience indicates that a single stage anoxic selector can provide excellent SVI control for most municipal wastewaters. A single zone size based on an F/M of about 1–2 kg COD/kg MLSS,day should provide excellent control of settleability. However, more efficient denitrification will be achieved by a staged configuration. If denitrification is also to be optimized, a 3-compartment selector with organic loadings based on the COD load to the first compartment as follows can be used:

First compartment   6 kg COD/kg MLSS,day
Second compartment   3 kg COD/kg MLSS,day
Third compartment   1.5 kg COD/kg MLSS,day

6. Because an anoxic selector removes soluble COD by denitrification, the DO concentration must be low. Thus, the selector contents must be mixed by mixers or by very low aeration rates and/or inefficient aeration devices. The activated sludge system must nitrify, and sufficient NO_3-N must be introduced into the selector either in the RAS stream or by internal mixed liquor recycle. If sufficient NO_3-N cannot be introduced into the selector to remove the necessary soluble COD, then the selector effect can be achieved by making only part of the selector anoxic. Typically, the initial portion of the selector would be anoxic and utilize the supplied NO_3-N; the downstream portion would be anaerobic due to the depletion of the supplied NO_3-N. Alternatively, if the first part of the selector is anoxic, then the second part can be aerobic (e.g., 2 anoxic compartments and 2 aerobic compartments). In another option, if the first part of the selector is anaerobic (i.e., it contains only RAS which is very low in NO_3-N and influent wastewater), then the second part can be anoxic (i.e., mixed liquor recycle could be introduced into these compartments). Due to variations in the relative proportion of COD which is oxidized and stored, values for the amount of soluble COD that can be removed by 1 mg NO_3-N/L vary considerably. For influent domestic wastewater, a typical value is 8 mg soluble COD/mg NO_3-N if little or no storage is occurring (i.e., essentially all soluble COD is being metabolized). Higher values will be observed if significant storage is occurring.

7. The overall sizing of an anoxic selector must allow sufficient time for denitrification to take place to a degree that reduces the soluble COD to the target selector effluent value (e.g., 60 mg soluble COD/L). Denitrification rates for the readily metabolizable soluble component of influent domestic wastewater—the material required to be removed in a selector—are typically on the order of 5–10 mg NO_3-N/g MLSS,hr at 20°C. Note that the objective in an anoxic selector is not the complete removal of NO_3-N; rather, it is the use of NO_3-N to remove soluble COD down to a level of approximately 60 mg/L. If denitrification is an additional objective to bulking control, a greater fraction of the initial zone of the aeration basin may have to be anoxic. Staging will also enhance the overall nitrogen removal efficiency, as discussed above.

8. Anaerobic selectors utilize both kinetic and metabolic selection processes. The overall sizing is dictated by the rate at which soluble organic matter is taken up (and orthophosphate is

released) under anaerobic conditions. For domestic wastewaters, the total anaerobic zone hydraulic retention time is usually in the range of 0.75 to 2.0 hours. This zone may be divided into compartments with the same $(F/M)_i$ values as the anoxic selector. For a true anaerobic selector, DO and NO_3-N must be excluded. For a nitrifying activated sludge this may require an anoxic zone at some location in the flow scheme.

9. Conditions in the downstream aeration basin will also affect sludge settleability. SVI will be optimized if plug flow conditions are achieved in the downstream aeration. Maintenance of fully aerobic conditions is also desirable.

10. The general criteria presented above may be used when site-specific data are not available. However, specific experience and data will allow refinement of the selector design and the performance achieved. These criteria for selector design will be illustrated by design examples and case histories.

Leopoldsdorf Sugar Mill

In Czechoslovakia and Austria, the following design criteria are used for aerobic selector design (Chudoba and Wanner, 1987)

Number of compartments	4 equal sized
Total selector F/M	3 kg BOD_5/kg MLSS,day
Initial compartment	
F/M $(F/M)_i$	12 kg BOD_5/kg MLSS,day
Selector oxygenation capacity	4 kg O_2/m^3,day
Total system F/M	0.3 kg BOD_5/kg MLSS,day
MLSS	3300 mg/L
Total system	
oxygenation capacity	2 kg O_2/m^3,day

Using these design criteria, Kroiss (1985) designed an aerobic selector for the Leopoldsdorf beet sugar mill, Marchfield, Austria, and obtained the results shown in Table 31 during the 1984 season. This activated sludge system had been previously plagued with severe bulking problems due to type 021N. The selector system had an $(F/M)_i$ of approximately 4.0 lb BOD_5/lb MLSS,day (overall F/M of about 0.10 to 0.15 lb BOD_5/lb MLSS,day). The selector removed 72% of the total COD, only 8.5% of which could be accounted for by oxygen uptake. SVI values of 50 mL/g were obtained at MLSS concentrations of 10,000 mg/L.

City of Hamilton, Ohio

The city of Hamilton, Ohio, activated sludge plant is a 25 MGD facility treating 19–20 MGD of wastewater consisting of approximately 7.5 MGD

Table 31. Technological Parameters of the Selector-Activated Sludge Process in the Leopoldsdorf Sugar Mill and the Results Obtained During the 1984 Season (Kroiss, 1985)

Parameter	Units	Value
Waste flow	MGD	12
Selector volume	MG	0.1
Aeration basin volume		4.3
Selector hydraulic residence time	hr	0.2
Aeration basin hydraulic residence time	hr	8.7
MLSS	mg/L	10,000
MCRT	day	8
SVI	mL/g	50
COD removal in selector	% of COD_{tot} removed	72
Oxygen consumed in selector	% of total oxygen consumed	8.5
Influent total COD	mg/L	450
Effluent soluble COD	mg/L	45
Enfluent total BOD	mg/L	340
Effluent soluble BOD	mg/L	10
COD removal efficiency	%	90
BOD removal efficiency	%	97
F/M	lb BOD_5/lb MLSS,day	~0.1
$(F/M)_i$	lb BOD_5/lb MLSS,day	~4

of domestic wastewater, with the remainder being paper manufacturing wastewater. Following primary clarification, the primary effluent is split between (1) a plant constructed in 1958 (T3), consisting of an aerated channel with a hydraulic retention time of approximately 8–12 min in which a mixed RAS and primary effluent stream flows prior to entering three diffused air aeration basins, and (2) a plant constructed in 1977, consisting of four single pass mechanically aerated aeration basins (T2A and T2B) (Figure 86). These basins were designed to operate as conventional, step feed, or completely mixed with separate introduction of RAS.

In 1982, when study of the Hamilton plant was initiated (Wheeler et al., 1984), the SVI of the T3 system was usually in the range 80 to 150 mL/g. In the T2A and T2B systems, SVIs below 200 mL/g were rarely encountered. The bulking was largely due to the filamentous organisms type 1851, *N. limicola*, type 0675, and type 0041. During late 1982 the T2A and T2B systems were repiped so that primary effluent and RAS flowed together in the step-feed channel prior to entering the aeration basin. This channel was aerated gently, and provided an average hydraulic retention time of approximately 7 min at average sewage flow. Figure 87 shows that following selector installation the SVI decreased gradually over an approximate seven-month period, and by mid-1983 had reached the same SVI range experienced by the T3 system. Following this, over the ensuing 6 years two noteworthy events have taken place.

First, during 1985 the T3 system was taken out of service for rehabilitation. During this approximate 6-month period the entire sewage flow was treated in the T2A and T2B systems. Figure 87 (and, in more detail, Figure 88) shows that there was an increase in SVI during this period, suggesting that the selector (now with an average hydraulic retention time of approximately 5 min) was overloaded, and soluble organic matter was breaking through into the main aeration basin. Once the T3 system was brought back into service the SVI values gradually returned to their previous values. This event suggests that the T2 system selector was sized very close to the limit for providing effective soluble organic matter removal, and this may also be the explanation for the long period required initially for it to reduce SVI.

The second event was the failure (in May 1985) of the air compressor used to generate air for selector aeration. It was not replaced. Since the Hamilton plant nitrifies completely, the selector is now really an anoxic, rather than an aerobic selector. And as such, it would appear from Figure 87 that it not only functions well, but probably better than when it was aerated. The additional selective pressure of requiring a rapid denitrification rate may explain this.

Davenport, Iowa

Albertson (1990) reported results of 2½ years of work on the use of aerobic selectors for bulking sludge control at Davenport, Iowa. This plant has an aeration basin consisting of 8 completely-mixed, submerged turbine-aerated cells which can be operated as a completely-mixed system or in a series mode with from 2 to 8 compartments in series. Even though the plant was operated in a series mode with 8 compartments in series, sludge bulking incidents due to types 0675, 1701, and 0041 occurred (as well as *Nocardia* spp. being present), and the SVI reached values in excess of 300 mL/g. The bulking sludge limited the 40 MGD design peak flow plant to 20–25 MGD. In late 1987 the plant staff started to control DO levels in the first aeration stage (the initial contact zone or ICZ) while operating the aeration basin in 2 trains of 3 compartments each. Within one month the SVI decreased from over 300 mL/g to less than 100 mL/g when the ICZ was operated with very low aeration rate at near zero DO (Figure 89). Further studies during 1988 (Figure 89) showed that reinstating full aeration in the ICZ induced bulking again. When the ICZ was operated with no aeration, an SVI of 40–50 mL/g resulted and led to dispersed floc and elevated secondary effluent SS concentrations. To control the SVI to between 80 to 100 mL/g, in 1989 the practice of aerating the ICZ for between 4 to 12 hr/day was adopted. When the SVI falls below 80 mL/g, the ICZ aeration period is increased; when the SVI exceeds 100 mL/g, the ICZ aeration period is decreased. Using this mode of operation the plant is able to operate with peak flows of 42 MGD and produce a secondary effluent with $cBOD_5$ = 7.4 mg/L and SS = 13.5 mg/L (20-month average values). The $(F/M)_i$ is typically 0.64 to 1.2 kg BOD/kg MLSS, day.

Upper Occoquan Sewage Authority (UOSA), Virginia

The UOSA Regional Water Reclamation Plant is a 27 MGD facility with preliminary treatment, primary treatment, and single sludge nitrification. Extensive physical-chemical treatment of

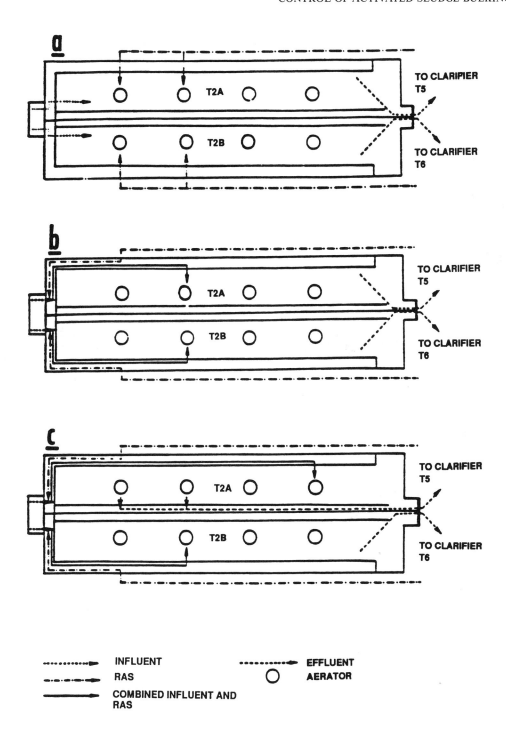

Figure 86. Modification of influent and RAS introduction points to aeration basins, Hamilton, OH: *a.* original system; *b.* modification of both systems to "selector" configuration using a selector channel hydraulic detention time of approx. 4 min.; and *c.* modification of system T2A to provide a selector channel hydraulic detention time of approx. 7 min. Reprinted with permission from Wheeler et al., "The use of a "selector" for bulking control at the Hamilton OH, USA, water pollution control facility," *Water Sci. Technol.*, 16, 1984. Pergamon Press plc.

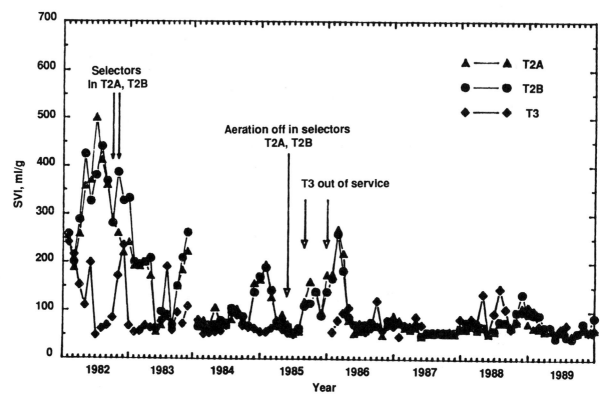

Figure 87. Monthly average SVI, Hamilton, OH.

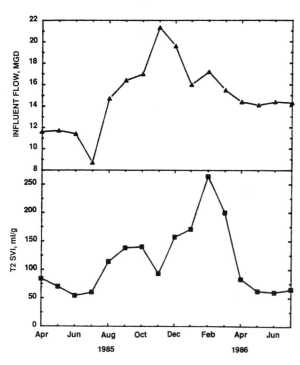

Figure 88. Effect of influent flow on T2 system SVI, Hamilton, OH.

the nitrified secondary effluent is provided to allow discharge of the treated effluent to a drinking water reservoir. Effluent discharge standards and plant reliability and redundancy requirements are extremely stringent, resulting in a conservative design for the biological treatment facilities with a relatively long hydraulic residence time, low organic loading rate, and low secondary clarifier overflow rate.

The biological treatment plant was upgraded and expanded from an existing completely-mixed activated sludge system which had historically experienced severe bulking problems with SVIs as high as 600 mL/g that restricted treatment capacity. Poor settleability resulted in reduced MLSS concentrations, reduced MCRT, and reduced capacity to nitrify. Periodic identification of the filamentous organisms present indicated that *M. parvicella* was typically the predominant filament, especially during cold weather. A bench-scale evaluation was conducted of a high DO aerobic selector (DO equal to or greater than 2 mg/L) for correction of the filamentous bulking problem. A full-scale aerobic selector was subsequently incorporated into the design of the expanded activated sludge system.

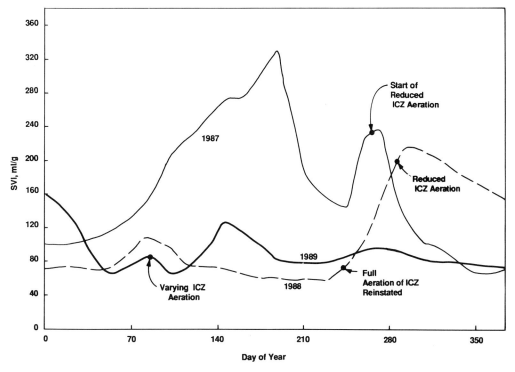

Figure 89. Bulking control at Davenport, IA.

Figure 90 is a schematic of the expanded, full-scale, upgraded system. The selector consists of an aerated channel (length:width ratio of 5:1) with a hydraulic residence time of 11 min based on influent flow. Dye testing indicated that the hydraulic flow pattern in the selector could be approximated as 3 equal-sized completely-mixed tanks in series. Following the selector are 2 sets of activated sludge aeration basins in series, the first being the newly added diffused air basins required to accomplish the plant expansion, and the second being the existing completely-mixed basins using slow-speed, surface mechanical aerators. The volumes of the new diffused air and existing mechanical aeration basins are approximately the same.

Good sludge settleability, as indicated by an average SVI of 74 mL/g, was obtained with the

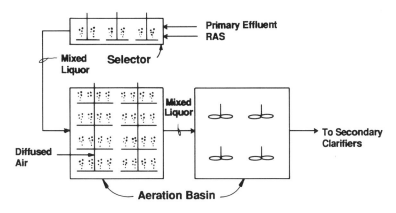

Figure 90. UOSA aeration basin and selector system. (Daigger and Nicholson, 1990). Reprinted by permission of the Water Environment Federation.

expanded and upgraded UOSA secondary treatment system while operating in excess of design organic loading and with periods where the hydraulic loading was equal to the maximum month design value. Figure 91 presents 4 years of data which illustrate the impact of the aerobic selector on sludge settleability. Prior to Mar 1988, the existing completely-mixed basins using surface mechanical aeration were in service. Severe sludge bulking problems were experienced regularly, as evidenced by SVI values typically greater than 150 mL/g and reaching 600 mL/g. Severe bulking conditions were experienced in early 1988 with SVI values increasing rapidly to nearly 500 mL/g. The predominant filament was *M. parvicella*. In Mar 1988 the high DO aerobic selector and diffused air aeration basins were brought on line and the existing completely-mixed basins were taken off line. The aeration basin hydraulic residence time remained about the same, since the volumes of the complete mix and diffused air basins were about the same. The result of the change in reactor configuration was a rapid decrease in SVI (Figure 91) as *M. parvicella* was eliminated from the system. The SVIs at UOSA have remained low since installation of the selector, except for a brief period in late 1988 when construction-related problems resulted in unusual operating conditions.

Operating conditions within the selector were characterized in Feb 1989. Direct measurements indicated soluble BOD_5 removals of approximately 60% and soluble COD removals of approximately 45% through the selector. Oxygen

requirements exerted in the selector were measured using two techniques:

1. Measurement of the oxygen uptake rate of samples withdrawn from the selector, and
2. Direct calculation from measured air flow rates using measured DO concentrations and oxygen transfer efficiencies as determined using an off-gas analyzer.

Both techniques produced similar results and indicated an oxygen requirement in the selector of approximately 0.1 mg O_2/mg soluble BOD_5 removed. The DO concentrations in the selector have been maintained consistently above 2 mg/L, suggesting that DO is not limiting the rate of oxidation of carbonaceous matter. This relatively low oxygen requirement suggests that storage, rather than oxidation, is the predominant mechanism for the removal of soluble organic matter in the selector. The F/M for the entire selector is 4.9 kg BOD/kg MLSS,day while the $(F/M)_i$ for the first selector compartment was 14.8 kg BOD_5/kg MLSS,day.

Tri-City, Oregon

This plant is an illustration of the use of an anoxic selector. The Tri-City, Oregon Wastewater Treatment Plant is a 13.5 MGD advanced secondary wastewater treatment plant consisting of preliminary treatment, primary treatment, activated sludge, and effluent chlorination. Typical secondary treatment (effluent BOD_5 and SS, both 30

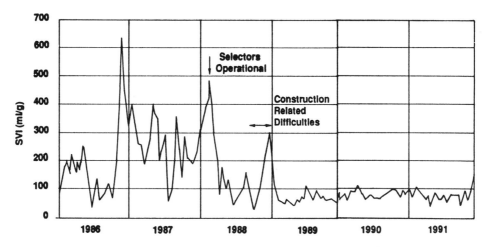

Figure 91. Effect of selector operation on SVI at UOSA. (Daigger and Nicholson, 1990). Reprinted by permission of the Water Environment Federation.

(mg/L) is provided during the winter, while advanced secondary treatment (effluent BOD_5 and SS, both 20 mg/L) is provided during the summer. Because the plant nitrifies during the summer and alkalinities are low, an anoxic selector was incorporated into the aeration basin both to reduce alkalinity consumption and to provide bulking control (Figure 92).

Diffused air aeration is provided throughout the basin; however, the initial 20% of the basin volume (the selector) is baffled from the rest (the aeration basin length-to-width ratio is 5:1), and both mechanical mixing and diffused air aeration equipment are provided in this section. Anoxic conditions are provided by operating the mixing but not the aeration equipment in the initial zone, and by returning nitrified mixed liquor from the effluent end of the aeration basin. Mixed liquor recycle pumping capacity equal to 100% of the design dry weather flow is provided. Fully aerobic treatment is provided by turning off the mixer and providing diffused air to the selector.

The data reported here are for anoxic selector operation over three dry seasons (Jun-Oct, 1986; Jun-Aug, 1987 and Oct, 1987; and May-Oct, 1988), during which an average SVI of 79 mL/g was obtained. Average selector hydraulic retention time and F/M are 86 min and 0.72, kg BOD_5/kg MLSS,day respectively.

The effectiveness of the selector was confirmed by operating the aeration basin in both the anoxic selector mode and in a fully aerobic step-feed mode. Figure 93 shows that the SVI increased whenever the aerobic, step-feed mode was used and decreased whenever the anoxic selector mode was used. Aeration of the selector, which was practiced in Aug and Sep 1987, also increased the SVI. The anoxic selector at Tri-City has some-times been too effective. It has on occasion reduced the SVI to as low as 20 to 30 mL/g and a turbid effluent has resulted. This has been countered by periodically providing a low level of aeration in the anoxic zone to encourage the growth of some low DO filamentous organisms and increase the SVI, as is also the practice at Davenport, Iowa.

Fayetteville, Arkansas

This is a case history of an anaerobic selector. The current Fayetteville, Arkansas, treatment plant commenced operation in Mar 1988 and replaced an existing completely-mixed aeration basin activated sludge system with a history of bulking sludge. The current facility was designed to nitrify and provide EBPR (supplemented by alum addition), as well as to provide bulking sludge control. The plant treats a mixture of domestic wastewater and food processing wastewater.

The aeration basin is configured so that it can be operated in both the A/O and the A^2/O modes; operation in the A^2/O mode is accomplished by using the mixed liquor recycle pumps. The A/O mode is typically operated. In this mode the anaerobic selector consists of 6 completely-mixed cells in series, followed by an aeration basin consisting of 4 completely-mixed cells in series with mechanical surface aerators (Figure 94). Mixed liquor recycle capability was also provided, but was not used during the evaluation period reported here.

From Jun 88 to Aug 89, the Fayetteville plant operated with a selector initial zone hydraulic residence time and $(F/M)_i$ of 15 min and 1.6 kg BOD_5/kg MLSS,day respectively; total aeration

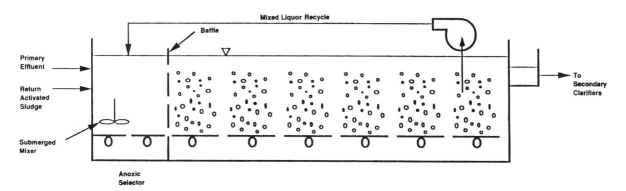

Figure 92. Configuration of anoxic selector system at Tri-City, OR. (Daigger and Nicholson, (1990). Reprinted by permission of the Water Environment Federation.

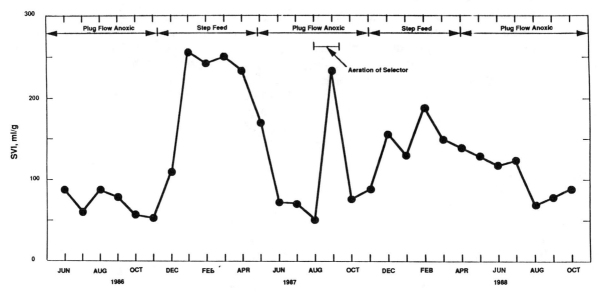

Figure 93. Performance of an anoxic selector at Tri-City, OR. (Daigger and Nicholson, (1990). Reprinted by permission of the Water Environment Federation.

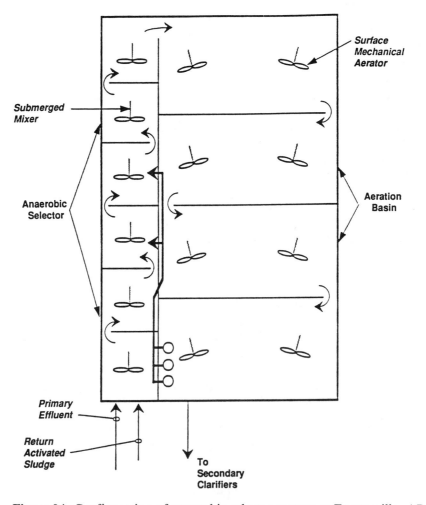

Figure 94. Configuration of anaerobic selector system at Fayetteville, AR.

basin hydraulic residence time was 9 hr and F/M was 0.045 kg BOD_5/kg MLSS,day.

Since start-up of the anaerobic selector system, the SVI at Fayetteville has averaged 86 mL/g and has not exceeded 185 mL/g. With the plant configuration prior to anaerobic selector installation, chlorination was used almost continuously to maintain the SVI below 150 mL/g. The highest SVI values have been observed during periods of rainfall when dilute, highly oxygenated wastewater caused the biological system influent to have a DO of between 8 to 10 mg/L. Low DO filaments apparently grow in the anaerobic selector during these periods.

23rd Avenue Plant, Phoenix, Arizona

This case study is discussed because it illustrates the ability to convert a high rate, non-nitrifying activated sludge plant with a bulking sludge to a nitrification-denitrification plant producing effluent with a total inorganic nitrogen of <7 mg/L, without additional aeration basin volume or air supply.

The 23rd Avenue plant was designed for 37 MGD, and was operated for many years at very low MCRT (0.8 to 1.3 days) and therefore very low MLSS concentrations (400–600 mg/L) to avoid *Nocardia* growth and foaming and partial nitrification. This mode of operation resulted in severe bulking sludge (SVI 300–600 mL/g) caused typically by the filamentous organisms type 1701, *S. natans, H. hydrossis, Thiothrix* spp., and type 021N. High rate operation also caused the oxygen transfer coefficient (α) to be very low; values on the order of 0.27 were estimated and O_2 transfer efficiency was approximately 6.9%.

High rate operation was conducted in the 4-pass aeration basins by using step feeding to passes 1 and 2 because of oxygen transfer limitations (Figure 95a) . In this mode the plant could treat only about 24 MGD.

During late 1989 to early 1990, the plant was converted by plant operations staff to a nitrification-denitrification plant incorporating an anoxic/oxic selector system. To accomplish this the plant was configured as a 4-pass aeration basin with the RAS and primary effluent fed to the head end of the first pass. Mixed liquor was recycled from the fourth pass of the aeration basin to the head end of the first pass. The first and second passes were compartmentalized (Figure 95b) to form a 5-compartment anoxic/aerobic selector system. The first 3 selector compartments were mixed by coarse bubble aeration;

the 4th selector compartment was provided with a combination of medium and coarse bubble aeration to allow either anoxic or oxic operation; the existing fine bubble diffusers were relocated from the first pass to the second pass (Figure 95b).

Figure 96 shows the dramatic improvement in operation and performance since installation of the selector system. The SVI has decreased from values that were often over 600 mL/g and routinely at about 300 mL/g to approximately 80–120 mL/g. This improvement in SVI has allowed the plant to operate at MLSS values on the order of 3000 mg/L (with RAS, SS of 9000 mg/L or more). The plant can now treat wastewater flows of approximately 33 MGD and has produced an effluent with the following average (15-months) constituent concentrations: $cBOD_5$ = 5.4 mg/L, SS = 6.6 mg/L, total inorganic N = 5.8 mg/L.

Converting to an anoxic/oxic selector system has improved the oxygen transfer situation by providing for conditions that increase the α value. In a high rate activated sludge system (MCRT <2 days), α values are suppressed most likely because of the presence of increased soluble organic matter in the mixed liquor. Reducing F/M (increasing MCRT) to provide a nitrifying activated sludge will increase α values, typically to the range 0.4 to 0.6. Combined with this effect, an anoxic selector provides an environment where soluble organic matter is removed in the absence of aeration. Therefore, the zone of high soluble organic matter, where oxygen transfer is inefficient, is made into a zone that does not require aeration. At the 23rd Avenue plant the combined effect of these factors resulted in the ability to treat a given BOD_5 load with the same amount of air (m^3 air supplied /kg BOD_5 removed) in the nitrification-denitrification (selector) mode and achieve total inorganic nitrogen removal to <7 mg/L, as was required for the high rate, non-nitrogen removal mode (Figure 96).

In the nitrification-denitrification anoxic selector mode, the fine bubble diffuser oxygen transfer efficiency increased to 12% with an average α value of 0.47; in the last two passes, the α value was 0.52.

Situations Where Selectors Are Not Effective

In the previous sections of this chapter, reference has been made to certain filamentous organism types against which selectors may not always be effective (Table 30). These are *M. parvicella*,

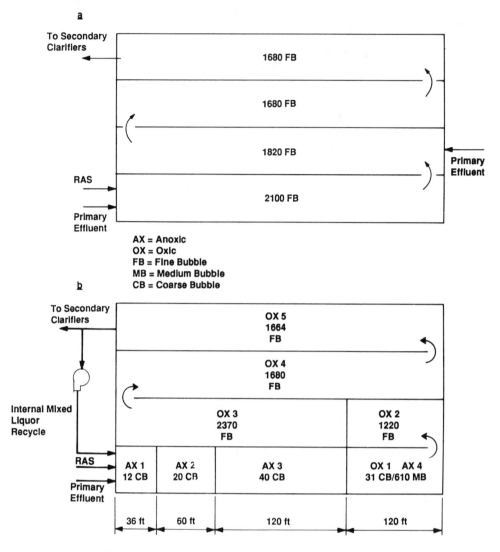

Figure 95. Modification of aeration basin to anoxic selector configuration at the Phoenix, AZ, 23rd Ave wastewater treatment plant.

type 0092, type 0675, and type 0041, especially when these organisms are growing in high MCRT (>15 days) BNR plants incorporating initial anaerobic and anoxic zones, and in high MCRT oxidation ditches with alternating aerobic/anoxic zones. The reasons why selectors are not always effective against these filamentous organisms in these types of situations is not completely understood. The hypotheses that have been proposed (but not proven) are that:

• These filamentous organisms grow on the degradation products of slowly biodegradable, soluble organics or of particulate organic matter, that are not removed in a selector. Dold and Marais (1986) suggest that the hydrolysis rate of particulate organics is zero under anaerobic conditions and about 40% of its aerobic value under anoxic conditions. Therefore, in activated sludge systems incorporating initial anaerobic and anoxic zones, particulate material is enmeshed in the activated sludge floc and transported to the main (aerobic) aeration basin, where its hydrolysis products provide the low soluble organic matter levels preferred by these filamentous organisms.

• Some of these filamentous organisms can denitrify and therefore can compete successfully in an anoxic selector

• Some of these filamentous organisms are low growth rate, low DO organisms. For

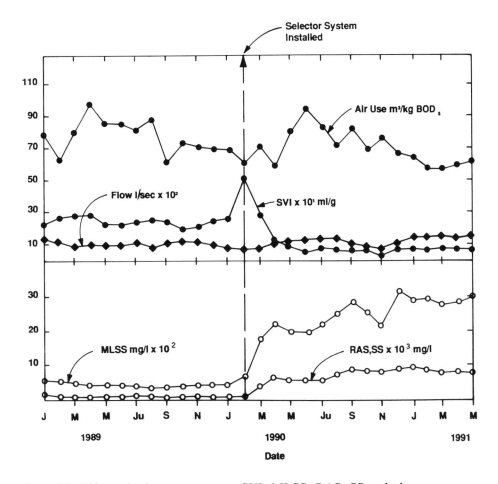

Figure 96. Effect of selector system on SVI, MLSS, RAS, SS and air use; Phoenix AZ, 23rd Ave. plant.

example, *M. parvicella* has been shown to grow well at low DO (Slijkhuis and Deinema, 1988).

Gabb et al. (1991) showed that these filamentous organisms (especially *M. parvicella* and type 0092) do not grow in high MCRT activated sludges that are fully aerobic (aerated all the time), whether or not the systems have a selector. When high MCRT, single reactor, continuous-flow laboratory activated sludge systems were intermittently aerated on a cycle of 6–7 min anoxic and 3–4 min aerobic, *M. parvicella* and type 0092 proliferated. When the aeration pattern was changed to continuous, the SVI values fell and *M. parvicella* and type 0092 were eliminated. Installation of an aerobic selector ahead of an intermittently-aerated aeration basin did not control the SVI. It is possible that the failure of aero-

bic selectors to control *M. parvicella* bulking at Tulsa, Oklahoma [Daigger and Nicholson (1990)] and at UOSA, Virginia [Daigger et al. (1985)] is because of the intermittent nature of the aeration in both of these systems. The aeration basins in these plants were both mechanically aerated and therefore most likely contained significant zones of low or zero DO.

Wanner and Novak (1990) worked with laboratory activated sludge systems with initial anaerobic zones, followed by either a completely-mixed or a compartmentalized aeration basin and a synthetic waste composed of glucose, ethanol, and acetic acid (soluble fraction) and a hard boiled egg suspension and peptone (particulate fraction). Even though the anaerobic selector removed the soluble organic matter, the filamentous organisms type 021N and *S. natans* grew in these systems. In systems with the compartmentalized aeration basins, the

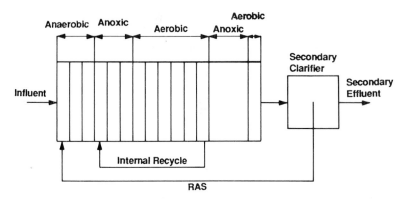

Figure 97. Schematic outline of the completely compartmentalized activated sludge system for bulking control and nutrient removal. From *Encyclopedia of Environmental Control Technology*, by Paul N. Cheremisinoff. Copyright 1989 by Gulf Publishing Company, Houston, TX. Used with permission. All rights reserved.

filamentous organism growth was less severe. These data suggest that transport of particulates through the anaerobic selector and their hydrolysis in the aerobic aeration basin can lead to filamentous organism growth. The experiments are not directly applicable to field observations of this type of bulking because the types of filamentous organisms observed are different — perhaps this was a function of differences in the substrate used compared to municipal wastewater.

Kucman (1987) conducted laboratory-activated sludge experiments on a synthetic wastewater based on glucose and ethanol in a system consisting of a completely mixed anoxic zone followed by a completely mixed aerobic zone; the MCRT was 10 days. The system was started with a sludge that was bulking due to type 021N. After a few days the type 021N had disappeared and the SVI decreased; however, the SVI gradually increased due to the presence of *M. parvicella*. No leakage of soluble degradable COD occurred from the anoxic zone; therefore, it was concluded that either *M. parvicella* can take up soluble substrate

under anoxic conditions or it can grow aerobically on hydrolysis products of the particulate material in the aeration basin. Since these experiments employed soluble substrate, the only particulate material available would be the biomass itself. Kucman (1987) demonstrated that the activated sludge dominated by *M. parvicella* had high denitrification rates and that its soluble substrate (glucose) uptake rates were of a magnitude that indicated substrate storage. These data were taken to indicate that *M. parvicella*, unlike other filamentous organisms, is able to denitrify and has a high unbalanced growth substrate uptake rate and storage capacity.

To combat the problems of soluble substrate leakage through anaerobic and anoxic zones into the aerobic zone and the generation of soluble substrate in the aerobic zone as a result of particulate matter hydrolysis, Wanner and Grau (1988) suggest compartmentalization of not only the initial anaerobic and anoxic zones, but also the aerobic zone. A flow sheet that incorporates this concept into a BNR plant has been proposed by Chudoba (1989) (Figure 97).

CHAPTER 4

Activated Sludge Foaming

TYPES OF ACTIVATED SLUDGE FOAM AND SCUM

The formation of foams or scums on the surfaces of activated sludge aeration basins and secondary clarifiers has been ascribed to a variety of causes. White, frothy foam is observed during the start-up (usually after 3–4 days) of activated sludge plants, possibly due to the presence of undegraded surface-active organic matter in the aeration basin. This type of foam usually disappears when the process becomes established. When poorly biodegradable, branched-chain alkyl benzene sulfonate-based household detergents were being marketed, persistent, voluminous, frothy, white foams were produced on aeration basins. Since the introduction of biodegradable household detergents, this type of foam does not occur in domestic wastewater treatment plants. Foaming due to the presence of slowly degradable surfactants occurs during the treatment of some industrial wastewaters. A sticky, viscous foam can occur when activated sludge is nutrient-limited, probably due to the formation of surface-active extracellular polymeric material by the activated sludge microorganisms. Foaming or scum formation, in conjunction with the production of a turbid effluent, can result from the excessive recycle of fine solids from solids handling processes (e.g., anaerobic digester supernatant, vacuum filter or belt press filtrate, and centrifuge centrate). This type of foam has a "volcanic" or "pumice-like" appearance. Microscopic examination will reveal that it is composed of amorphous particles. Scum can be produced on secondary clarifiers or in the anoxic zones of nitrifying activated sludge plants by denitrification. Small bubbles of nitrogen gas attach to the activated sludge flocs and carry them to the surface. When occurring in secondary clarifiers this phenomenon often is referred to as "sludge-rising" or "blanket-rising." A viscous, stable and often chocolate-colored foam or scum has been noted widely on activated sludge aeration basins and secondary clarifiers (Anon., 1969; Wells and Garrett, 1971; Lechevalier, 1975; Pipes, 1978b; Dhaliwal, 1979). The occurrence of this

foam has been associated with the presence in the activated sludge of large numbers of bacteria of the genus *Nocardia*. A similar type of foam has been observed in the presence of the filamentous organism *M. parvicella* (Seviour et al., 1990). It is this type of foam that will be discussed in this chapter.

EXTENT AND SIGNIFICANCE OF *NOCARDIA* FOAMING PROBLEMS

Surveys of *Nocardia* spp. incidence in activated sludge have been conducted by Richard et al. (1982a), Strom and Jenkins (1984), and Blackbeard et al. (1986) in conjunction with their surveys of filamentous organisms in bulking activated sludge. *Nocardia* spp. was one of the most commonly observed filamentous organisms in activated sludge, being present as a dominant filamentous organism in approximately 30% of the USA samples examined and in 14% of the South African samples (Table 19).

Some surveys have been directed specifically to foaming-activated sludges or to *Nocardia* incidence in activated sludge. Seviour et al. (1990) surveyed 129 activated sludge plants in Queensland, New South Wales and Victoria, Australia, and found that 66 of these plants had a foam problem at the time of the survey. The most commonly found filamentous organisms concentrated in the foam over their levels in the mixed liquor were *M. parvicella*, *N. amarae* (Figures 30a, b, and c, and Figures 98a, b, and c) and the so-called pine-tree-like organism (*N. pinensis*) (Blackall et al., 1989) (Figure 15b, and Figures 98d, e, and f). The foams also contained other filamentous organisms such as types 0092, 0914, 0041, and 0675, but these were present at concentrations no greater than in the mixed liquor.

Pitt and Jenkins (1990) conducted a nationwide survey of activated sludge foaming in the USA. Of the 114 responses, 75 (66%) indicated that they had experienced some type of foaming at one time or another. As part of this survey the *Nocardia* levels in foam and mixed liquor

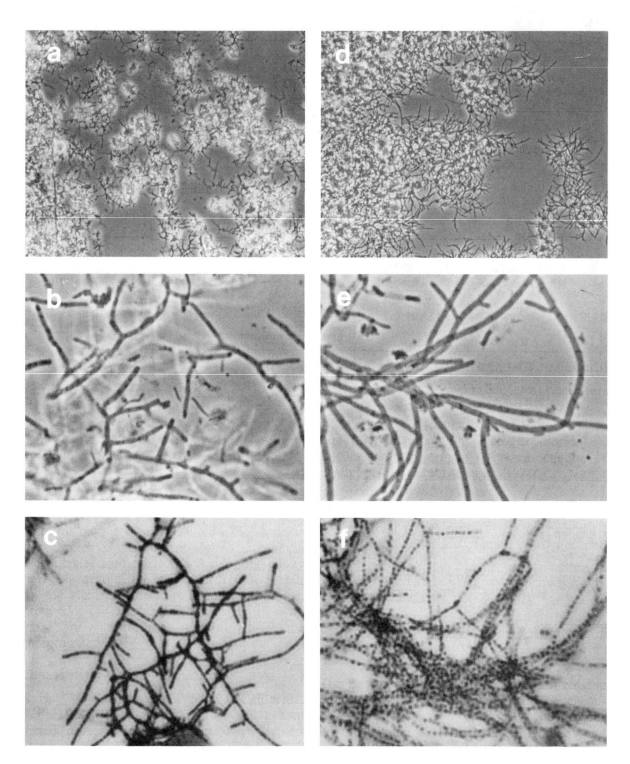

Figure 98. Microscopic appearance of *Nocardia* foam: *a.*, *b.* and *c.*, *N. amarae*; *d.*, *e.* and *f.*, *N. pinensis* (*a.* and *d.* 100X; *b.* and *e.* 1000X; *c.* and *f.* Gram stained, 1000X), (*a.-e.* phase contrast; *f.* transmitted light).

samples were determined for the plants in terms of a numerical scale rating based on the subjective scoring described in Table 14. The average numerical scale rating for plants indicating a foaming problem was 4.2 ("common") compared to 2.1 ("few") for those not indicating a foaming problem. *N. amarae* was determined by Lechevalier and Lechevalier (1974) to be the most common actinomycete present in activated sludge foam in the USA. This species has not been reported in environments other than activated sludge. Lechevalier (1975) also regularly isolated *N. rhodochrus Rhodococcus* spp. *N. asteroides*, *N. caviae*, and strains of *Mycobacterium spp.* from activated sludge foams. Members of the *N. rhodochrus* group have been isolated more commonly from foaming activated sludge in West Germany and Switzerland (Lemmer and Kroppenstedt, 1984) than in the USA.

Some of the actinomycetes that have been isolated from activated sludge foams are opportunistic human pathogens (e.g., *N. caviae*, *N. brasiliensis*, *N. asteroides*, and strains of *Mycobacterium*) (Lechevalier, 1975). Although no incidents of human infection with *Nocardia* from wastewater treatment plants have been reported, the potential exists that aerosols generated by aeration systems can disperse these organisms to where they come in contact with treatment plant workers or inhabitants of the area surrounding wastewater treatment plants. The significance of this potential should not be underestimated, especially with the rise in the numbers of immunologically-suppressed individuals in the population.

The presence of significant quantities of a *Nocardia*-type foam can cause severe operating problems. In aeration basins with submerged effluent structures, foam tends to accumulate in the basins (Figures 99a, b, and c). It can accumulate to such an extent that it overflows the basin freeboard and covers walkways, handrails, and surrounding areas, creating hazardous, slippery conditions and preventing access to appurtenances (Figure 99d). In covered aeration basins *Nocardia*-type foams have been known to accumulate to such a degree that the water content of the foam trapped in the basin exceeds the available head for gravity flow of sewage through the aeration basin thereby preventing the influent wastewater from entering the basin. Process control becomes extremely difficult if a significant fraction of the solids inventory is present in the foam trapped in the aeration basin. In one plant it was estimated that the foam contained 40% to 45% of the total activated sludge suspended solids inventory. Hao et al. (1988) estimated that for a 4.5 m deep aeration basin with an MLSS concentration of 3000 mg/L and a 7.5 cm deep foam containing 5% TSS, some 20% of the total sludge inventory would be in the foam. When foam escapes from the aeration basin into the secondary clarifier it may overwhelm surface scraping devices (Figure 99e) and enter the secondary effluent, elevating effluent SS levels. In cold climates *Nocardia*-type foam can freeze on the secondary clarifier surfaces, putting scum removal devices totally out of commission (Figure 99f). In warm climates the accumulated foam rapidly becomes odorous.

A survey of USA plants with anaerobic digester foaming problems by van Niekerk et al. (1987) showed that many such digester foaming incidents can be attributed to the presence of *Nocardia* in the waste activated sludge fed to the digester. van Niekerk et al. (1987) showed in laboratory-scale anaerobic digester experiments that digesters fed with primary sludge alone did not produce stable foam, but that when waste-activated sludge containing *Nocardia* was digested with the primary sludge a stable foam was produced. Anaerobic digester foaming due to *Nocardia* has caused blockage of gas mixing devices, inversion of digester solids profiles, gas binding of sludge recirculation pumps leading to an inability to heat the digesters, entrapment in and fouling of gas collection pipework, foam penetration between floating covers and digester walls, and tipping of floating covers by rapid expansion and then collapse of the sludge volume when gas mixers are operated intermittently.

THE *NOCARDIA* FOAMING PROBLEM

Two different but interrelated aspects must be addressed to understand *Nocardia* foaming. These are the factors that encourage the growth of *Nocardia* in activated sludge, and the factors that encourage foaming. Although interrelated, the former factors are largely microbiological and the latter, mainly physical/chemical. The foam-causing factors will be addressed first.

A foam is a dispersion of gas (air) bubbles in a liquid (water) or a solid (e.g., styrofoam). It is a 2-phase system. For air-in-water foams, the air is the dispersed phase and water is the continuous phase (Figure 100a). Foams collapse by the drainage of water by gravity so that eventually the

Figure 99. Nocardial foaming in activated sludge: *a.*, *b.* and *c.* aeration basin foaming; *d.* a slippery walkway; *e.* an overwhelmed scum trap; *f.* frozen nocardial foam.

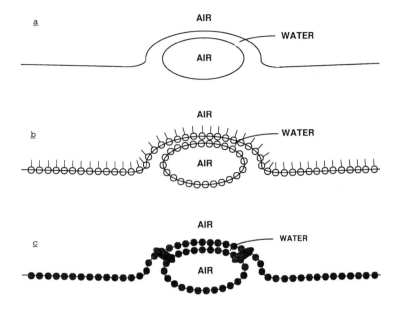

Figure 100. Air-in-water foams of various types: *a*. No surface active materials present. *b*. Surface active materials present that accumulate at air-water interface and stabilize it. *c*. Hydrophobic solids present that accumulate at interface and stabilize it; note also that some solid particles are large enough to bridge across the air bubble and prevent drainage of water from it.

water layer separating two air bubbles thins to the point where the bubbles burst. Foams can be stabilized by surface active agents (surfactants), which are molecules that are attracted to the air-water interface (Figure 100b). In a sense these molecules toughen up the water film between the air bubbles and allow it to become much thinner before it ruptures. Foams containing surfactants therefore last longer than foams without them.

Another way in which foams can be stabilized is by the addition of hydrophobic particles (particles that are poorly wetted by water). This produces a 3-phase foam consisting of dispersed phases of air and the hydrophobic solid and a continuous water phase. The hydrophobic solid particles tend to congregate at the air-water interface, again making it stronger. If the hydrophobic particles are large enough they may bridge the water film between two air bubbles and create a dam that prevents the liquid drainage and film thinning that eventually results in film rupture. (Figure 100c).

Nocardia spp. have hydrophobic cell walls due to the presence of long-chain mycolic acids on their surfaces (Minnikin, 1982). If *Nocardia* grows in sufficient numbers in activated sludge, they will render the flocs hydrophobic enough

that they will attach to air-water interfaces (e.g., bubbles in an aeration basin) and be carried to the basin surface. At the surface the foam will drain and the scum or foam will become more concentrated in terms of SS and *Nocardia* content than the mixed liquor. SS concentrations in *Nocardia* foams on aeration basins as high as 4% to 6% have been observed. Pitt and Jenkins (1990), in their survey of USA activated sludge plants, found that *Nocardia* foams always had higher numerical scale ratings than the mixed liquors from which they were derived. Wheeler and Rule (1980) reported that a mixed liquor contained up to 10^6 *Nocardia* microcolonies/mL and foam generated from it contained up to 10^{12} *Nocardia* microcolonies/mL. Vega-Rodriquez (1983) observed that when activated sludge containing *Nocardia* was aerated in the laboratory, typically 75% to 90% of the *Nocardia* filaments (on a filament length basis) were transferred into the foam.

The presence of surfactants enhances both the amount of foam produced by a *Nocardia*-containing activated sludge as well as increasing the foam stability. For surfactants to exert this effect they must be poorly or slowly biodegradable so that they persist in the aeration basin. Ho and Jenkins (1991) demonstrated this effect by conducting foaming experiments on mixtures of various amounts of a slowly biodegradable non-ionic surfactant—the alkyl phenol ethoxylate, Igepal C-620, using: (a) an activated sludge that contained *Nocardia* (from the Victor Valley, California, activated sludge system), and (b) an activated sludge that did not contain *Nocardia* [from the East Bay Municipal Utilities District (EBMUD), Oakland, California, activated sludge system] (Figures 101 and 102). In the Victor Valley sludge there was considerable transport of SS out of the activated sludge during foaming, and the foam height increased significantly as the surfactant concentration increased. A stable brown SS-containing foam was produced. In the EBMUD-activated sludge the SS content of the activated sludge did not decrease significantly during foaming. A small amount of very unstable white foam containing little SS was produced. Foaming of *Nocardia*-containing activated sludge reduced the *Nocardia* level in the activated sludge by almost 3-fold.

These observations suggest that a useful analogy can be drawn between *Nocardia* foaming in activated sludge and the process of mineral beneficiation by flotation (Willis, 1988). In mineral beneficiation, minerals of commercial value are separated from nonvaluable material (gangue) by flotation. A chemical called a collector is added

to the ground-up mixture of mineral, gangue, and water that specifically absorbs on the mineral particles and renders them hydrophobic. A surfactant is also added. The hydrophobic mineral-collector combination attaches to an air bubble and is floated to the surface. The surfactant serves to stabilize the float so that it can be skimmed off.

In activated sludge, *Nocardia* filaments can be regarded as the hydrophobic collector for activated sludge flocs. The aeration air produces a float of *Nocardia*-containing activated sludge flocs that is stabilized either by the presence of poorly degradable surfactants in the influent or by biologically produced surfactants. In this sense it is better to view *Nocardia* scum formation as a flotation process rather than a foaming process.

In many instances of *Nocardia* foaming in activated sludge plants, anecdotal information on the "history" of the foaming problem often reveals the presence of modest amounts of foam for much of the time, accompanied by very severe foaming incidents with a rapid onset (e.g., in a matter of hours). It is possible that these severe foaming events may be related to increased surfactant levels. Thus, one could reason that a *Nocardia*-containing activated sludge in the absence of surfactant could produce a modest amount of foam. A slug discharge of poorly bio-

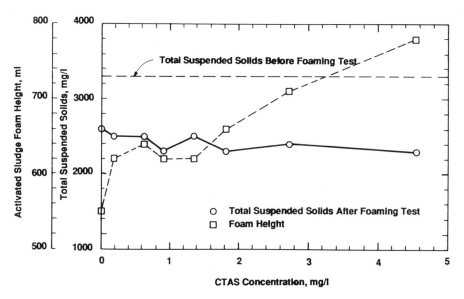

Figure 101. Effect of foaming on total solids of Victor Valley RAS (*Nocardia*-present) containing Igepal C-620. Reprinted with permission from Ho, C.-F. and Jenkins, D., "The effect of surfactants on *Nocardia* foaming in activated sludge," *Water Sci. Technol.*, 23, 1991. Pergamon Press plc.

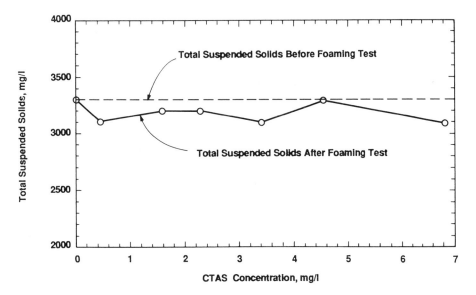

Figure 102. Effect of foaming on total solids of EBMUD RAS (*Nocardia*-free) containing Igepal C-620. Reprinted with permission from Ho, C.-F., and Jenkins, D. "The effect of surfactants on *Nocardia* foaming in activated sludge," *Water Sci. Technol.*, 23, 1991. Pergamon Press plc.

degradable surfactant (e.g., from clean-up processes in an industry) to this system could provide the foam stabilization that would cause a rapid onset of severe *Nocardia* foaming in the wastewater treatment plant.

One prerequisite for the production of a foam or a float is the presence of a gas phase. Since all activated sludge systems are aerated in one way or another, it is not possible to eliminate the gas phase from an aeration basin. This approach, however, is a viable method for reducing *Nocardia* foaming in anaerobic digesters. All anaerobic digesters produce gas from the digestion process; however, a far greater volume of gas is passed through the digesting sludge by gas mixing devices than the volume of gas produced by digestion. van Niekerk et al. (1987) showed that the foam level produced in laboratory anaerobic digesters being fed a mixture of primary sludge and waste activated sludge containing *Nocardia* was a function of the digester mixing method. The most foam was produced by fine bubble gas mixing, followed by coarse bubble gas mixing, followed by mechanical mixing (Figure 103). These results suggest that *Nocardia* foaming in digesters may be minimized through the use of a mixing system which does not introduce small to medium size bubbles into the digester.

Nocardia foam problems are increased enor-

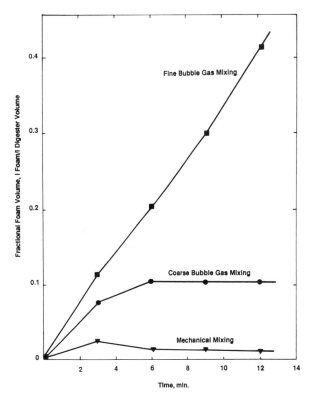

Figure 103. Foam accumulation rate in experimental anaerobic digesters as a function of mixing technique (van Niekerk et al., (1987). Reprinted by permission of the Water Environment Federation.

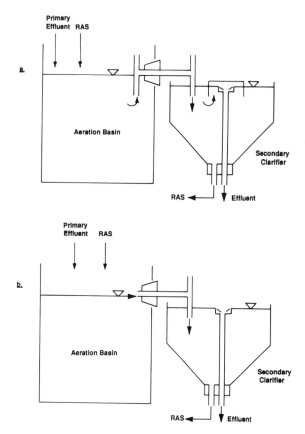

Figure 104. Detail of aeration basin outlet and secondary clarifier overflow for: *a.* foam "trapping" and; *b.* "non-trapping" units. (Cha et al., 1992). Reprinted by permission of the Water Environment Federation.

mously when the floated material is trapped on the surface of treatment units, and especially when the trapped foam is removed and recycled through the treatment plant. Foam trapping occurs whenever the free water surface is intercepted by a structure (e.g., a wall or baffle) that makes the water leave the treatment unit by a subsurface route rather than by an overflow.

Cha et al. (1992) compared the *Nocardia* populations in laboratory-activated sludge systems with the aeration basin/secondary clarifier configurations illustrated in Figure 104. The trapping configuration had a subsurface aeration basin drawoff and a scum baffle in front of the secondary clarifier weir. The nontrapping configuration had an overflow aeration basin outlet and no secondary clarifier scum baffle. Figure 105 shows that the *Nocardia* populations in the "trapping" unit were up to 5 times those in the "nontrapping" unit. It should be noted that this *Nocardia* population increase was due only to foam trapping, since Cha et al. wasted foam and mixed liquor in the same proportions. Foam recycle will increase *Nocardia* levels even more.

Based on the results of Cha et al. (1992), it is tempting to answer the often-posed question, "why do we have all this *Nocardia* foaming now in the USA when we never did before?" in the following way, "Until the secondary treatment requirements of the 1972 Clean Water Act, there was no requirement for placing scum baffles in

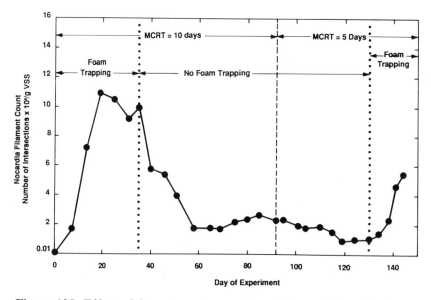

Figure 105. Effect of foam trapping on *Nocardia* populations in the bench-scale activated sludge units. (Cha et al., 1992). Reprinted by permission of the Water Environment Federation.

secondary clarifiers. Under these circumstances anything (including *Nocardia*) that floated in a secondary clarifier was washed out of the treatment plant in the secondary effluent. Therefore, to float was a disadvantage. With the installation of secondary clarifier scum baffles, floating material (including *Nocardia*-rich activated sludge particles) is now captured and often is recycled back through the treatment plant. Therefore, to float is a great advantage because it means retention in the biological system."

An additional effect of foam trapping was suggested by the results of pure culture chemostat experiments on *N. amarae* strains isolated from San Francisco and Sacramento, California, activated sludges (Blackall et al., 1991). The two chemostat configurations shown in Figure 106 produced significantly different *N. amarae* growth forms. When the chemostat culture effluent was drawn off by an overflow weir, the *N. amarae* culture grew largely as clumps with diameters of up to 500 μm, suspended in a relatively clear liquid. When the chemostat culture withdrawal system was changed to a subsurface withdrawal device (Koopman et al., 1980), the *N. amarae* grew almost completely in the form of dispersed filament fragments. These observations can be explained by the fact that the hydrophobic *Nocardia* filaments tend to float (especially if the culture is aerated with bubbles, but also when head space aeration is employed). With an overflow culture withdrawal, floating filaments will be selectively removed. If the filaments are able to aggregate and form clumps that are heavy enough to overcome their propensity to float, then these will be selectively retained in the chemostat. Hence, a clumped culture will develop in a chemostat with a surface overflow culture withdrawal. When the culture withdrawal is subsurface, floating filaments will not be selectively removed. A dispersed culture will be formed under these circumstances because growth in a dispersed form no longer offers the disadvantage of loss from the culture by floating; in addition, compared to growth as clumps, dispersed cells have better access to the substrate.

These data can be applied to observations made in activated sludge systems. When foam trapping occurs in activated sludge, free-floating *Nocardia* filaments are usually seen in the bulk liquid outside the activated sludge flocs. (In fact, their presence can be used to detect a foam trapping problem.) Since free-floating *Nocardia* filaments expose a greater amount of hydrophobic surface to the aqueous phase than if the filaments

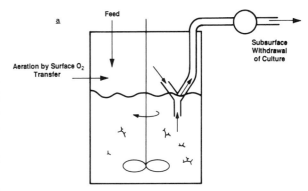

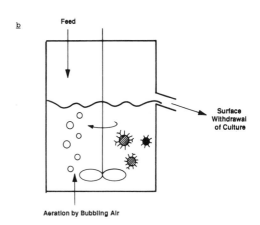

Figure 106. Chemostat configurations and their effect on *N. amarae* growth form: *a.* chemostat with head-space aeration and subsurface culture withdrawal (chemostat produces dispersed *N. amarae* growth); *b.* chemostat with standard bubble aeration and a surface overflow for culture withdrawal (chemostat produces clumped *N.amarae* growth). (Blackall et al., 1991). Reprinted by permission of the Water Environment Federation.

were inside the flocs (nonfoam-trapping situation), they will provide a greater enhancement of foaming. Therefore, foam trapping not only increases *Nocardia* levels in activated sludge but also produces a culture with more free *Nocardia* filaments and a greater propensity to foam.

Caution must be paid in applying the results of laboratory foaming tests to full-scale activated sludge systems. In a laboratory foaming test the maximum depth of liquid used is on the order of 1–2 ft. In an aeration basin, the liquid depth is usually on the order of 15 ft or more. In foaming, *Nocardia*-containing flocs are brought to the liquid surface. Because the column of liquid below a unit area of liquid surface is much greater in a full-scale aeration basin than in a laboratory foaming test, it will take less *Nocardia* to cause a stable foam in a full-scale aeration basin than in a laboratory foam test. Therefore, the typical laboratory foam test is an insensitive indicator of foaming in prototype aeration basins.

The previous discussions of *Nocardia* foam formation and the influences on it of various physical/chemical factors and of reactor design should serve to indicate that foaming tests are poor methods for estimating *Nocardia* populations in activated sludge—there are just too many factors involved in foaming other than the actual *Nocardia* population. For this reason, a convenient *Nocardia* filament counting method has been developed (Table 6) by Pitt and Jenkins (1990), based on the work of Vega-Rodriguez (1983). This technique can be replicated between observers of the same sample with a coefficient of variation of about 20%.

NOCARDIA GROWTH AND ITS CONTROL IN ACTIVATED SLUDGE

It is probably fair to state that, at the present, more is known about how to prevent *Nocardia* from growing in activated sludge than what causes it to grow!! Lemmer (1986) presented several postulates concerning the growth of scum-causing actinomycetes in activated sludge. Nocardias are able to use a wide variety of organic substrates as carbon sources. These include readily degradable compounds such as sugars and low molecular weight fatty acids as well as slowly biodegradable, high molecular weight compounds including polysaccharides, proteins, pesticides, and aromatic compounds (Lemmer and Kroppenstedt, 1984). Some Nocardias can grow

saprophytically on dead cell material (Segerer, 1984). The cell yield of Nocardias was found to be proportional to substrate concentration over a very wide range of substrate levels—from extremely low levels (several mg/L) up to 15,000 mg/L (Segerer, 1984). Lemmer (1986) postulated that Nocardias used these growth capabilities to produce the high populations associated with activated sludge foaming. She proposed that the ability to grow on slowly degradable complex organic matter provides Nocardias with a niche that other activated sludge heterotrophs are unable to occupy; growth on these substrates (at low concentrations) is slow and a small population of Nocardias result. When high concentrations of readily biodegradable substrate become available, it was proposed that the ability of Nocardias to produce cell material proportional to substrate concentration over a wide range of substrate concentrations allows the population of Nocardias to increase to the high levels characteristic of foaming activated sludge. Some doubt has been cast on the validity of this hypothesis by the pure culture chemostat data of Blackall et al. (1991) on the growth of two strains of *N. amarae* (ASF3 from San Francisco, CA activated sludge and ASAC 1 from Sacramento, CA activated sludge) on acetate. Table 32 shows that maximum growth rates of *N. amarae* strains were of the same order as those of floc-forming heterotrophic bacteria from a completely mixed activated sludge plant and for the filamentous organism type 021N. Compared to the floc-forming bacterium *Zoogloea ramigera* from a selector activated sludge system, the *N. amarae* strains were poor competitors both at low growth rates (low substrate concentrations) and at growth rates close to the respective maximum growth rates (high substrate concentrations) (Figure 107).

Lemmer (1986) further postulated that Nocar-

Table 32. Steady-State Data for *N. amarae* Strains, Type 021N, and *Z. ramigera* (Blackall et al., 1991)

Strain	μ_{max}, day^{-1}	K_{AC}, mg acetate/L	K_{DO}, mg O$_2$/L	Y, gSS/g acetate
N. amarae				
ASF3	2.8	0.49	0.13	0.42
ASAC1	3.0	2.5	—	0.41
type 021N	3.8	0.07	0.06	0.38
AS2[a]	2.5	—	—	0.36
Z. ramigera	5.5	0.30	0.12	0.41

[a]Average of floc formers from a completely mixed activated sludge system (van Niekerk et al., 1987).

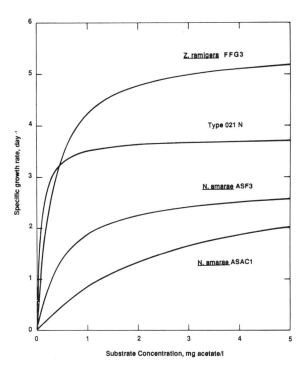

Figure 107. Predicted curves relating growth rate to acetate concentration for *N. amarae* strains, type 021N and *Z. ramigera*. (Blackall et al., 1991). Reprinted by permission of the Water Environment Federation.

dias, because of their hydrophobic properties and their propensity to float in aeration basins, have selective access to substrates that float. Such materials include oils, fats, and greases. Nocardias have often been observed adhering to, and even penetrating, oil droplets (Marshall and Cruickshank, 1973) and, when grown on hydrocarbons, they can produce biosurfactants that emulsify nonpolar substrates (Margaritis et al., 1979; MacDonald et al., 1981; Cairns et al., 1982) and may also help in concentrating substrates at the air-water interface. Further, Nocardias are adapted to dryness (Dommergues et al., 1978), and can protect themselves against solar radiation by producing pigments (Krinsky, 1979). Taking all of these factors into account suggests that a surface foam may provide an environment where Nocardias can compete well in terms of both substrate availability and from a survival standpoint.

In support of these postulates, Matsche (1980) and Pipes (1978b) have both reported that the discharge of wastes high in oil or grease to an activated sludge plant can cause a rapid increase in *Nocardia* populations. However, since there are many instances of *Nocardia* foaming in wastewater with low oil and grease content, these substrates cannot be the sole reason for excessive *Nocardia* growth in activated sludge.

Anecdotal information suggests that the presence of Nocardias in activated sludge is associated with high, rather than low, temperatures and with high MCRT values, rather than low MCRTs. Pipes (1978) suggested that *Nocardia* growth (and foam formation) in activated sludge was a problem associated with high MCRT (≥ 9 days) and warm temperatures ($>18°C$). Wilson et al. (1984) reported that successful *Nocardia* control by washout was achieved at the Phoenix, Arizona, 23rd Avenue treatment plant only when the MCRT was decreased to approximately 1 day. Summer wastewater temperatures at this plant reach 28°C. Beebe (1983) reported that *Nocardia* foaming in the first stage of the two-stage activated sludge system at the San Jose/Santa Clara, CA, Wastewater Treatment Plant could be controlled by increased sludge wasting (increased F/M, decreased MCRT), accompanied by chlorination of the RAS for a short period of time. Typical successful operating parameters for this technique were to increase F/M from 0.35 to 0.5 kg BOD_5/kg MLVSS, day, and to chlorinate the RAS at approximately 4 kg Cl_2/10kg MLVSS, day for a period of 4–5 days.

Pitt and Jenkins (1990) and Cha et al. (1992) conducted laboratory-scale activated sludge experiments to determine the relationship between *Nocardia* populations in activated sludge and MCRT for a range of temperatures. In both studies, completely mixed activated sludge units that employed the foam trapping features illustrated in Figure 104a were used. Pitt and Jenkins' experiments were done at the Southeast Water Pollution Control Plant in San Francisco, CA, while Cha et al.'s experiments were conducted at the Sacramento, CA Regional Wastewater Treatment Plant. The results of these studies are presented in Figure 108 and show that *Nocardia* populations generally increase with increasing MCRT over the entire range of MCRT tested (approximately 1.5 to 20 days). For the 16°C Sacramento reactor, *Nocardia* populations declined to undetectable levels ($<1 \times 10^4$ intersections/g VSS) at an MCRT of 2.2 days; for the 24°C Sacramento reactor, washout occurred at an MCRT of 1.5 days. In the San Francisco reactors, the same pattern of increasing *Nocardia* populations with increasing MCRT and temperature was observed; however, it was neither possible to obtain a wash-

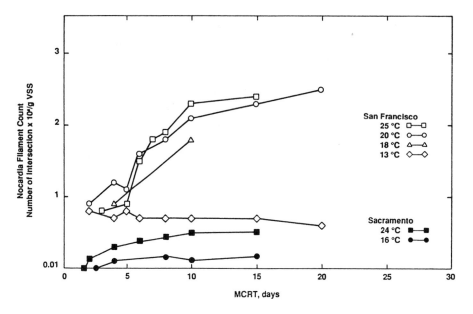

Figure 108. Comparison of *Nocardia* populations in bench-scale activated sludge units at the City of San Francisco (SEWPCP) and the County of Sacramento (SRWTP) at a range of MCRTs and temperatures. (Cha et al., 1992). Reprinted by permission of the Water Environment Federation.

out of *Nocardia* to undetectable levels nor to discern any difference, at temperatures of 18°C, 20°C, and 25°C, in the MCRT at which *Nocardia* populations began to decline from their levels at very high MCRT. Furthermore, all of the Sacramento *Nocardia* counts were below the level that was concluded to be the "no *Nocardia* growth" level (at 13°C) at San Francisco. It is suggested that the difference between these data sets was due to the different practices employed at the two plants for dealing with removed foam and the recycle streams from the solids handling processes (Figure 109). At the time of the study in San Francisco, the foam from the mixed liquor channels was returned to the headworks. At Sacramento, secondary scum from the mixed liquor channel was first sent to the WAS dissolved air flotation units, and thence to the anaerobic digesters. After anaerobic digestion, all of the digester contents were transferred to solids retention basins, which have a sludge residence time on the order of 5 years. Supernatant from the solids retention basins and subnatant from the dissolved air flotation units are both returned to the headworks. These differences suggest that seeding of the influent with *Nocardia* from removed and recycled foam, and from solids treatment process recycle streams is the cause of

both the much higher *Nocardia* populations in the San Francisco activated sludge, and of the inability to wash *Nocardia* out from the San Francisco activated sludge. Attempts to confirm this by detecting *Nocardia* filaments in the primary effluent entering the San Francisco activated sludge system were not successful (Pitt and Jenkins, 1990).

The MCRT values obtained for *Nocardia* washout in these studies agreed very well with those being employed in practice at the Sacramento and San Francisco plants. Table 33 presents these data, and Figure 110 shows an Arrhenius equation plot for a combination of the Sacramento pilot plant data and the full-scale plant data. These data can be utilized to determine the MCRT required to wash out *Nocardia* from activated sludge over a range of temperatures.

Cha et al. (1992) examined the effect of pH on *Nocardia* populations in activated sludge using completely mixed laboratory activated sludge units operated over a range of pH values between 6.0 and 7.5, and at MCRT values of 3 and 8 days. Figure 111 shows that at both MCRTs, pH influenced the *Nocardia* counts in the activated sludge. This influence was greater at the 8-day MCRT and indicated an optimum in the region

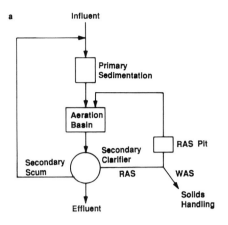

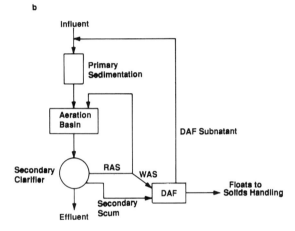

Figure 109. Schematic drawing of the flow of scum and recycle streams at: *a.* San Francisco Southeast Water Pollution Control Plant; and *b.* Sacramento Regional Wastewater Treatment Plant. (Cha et al., 1992). Reprinted by permission of the Water Environment Federation.

of pH 6.5; a similar trend was evident in the data at an MCRT of 3 days. The value of the pH optimum is quite significant from the viewpoint of *Nocardia* occurrence in full-scale activated

sludge systems. In their survey of *Nocardia* foaming in prototype activated sludge plants, Pitt and Jenkins (1990) found that every one of the 6 oxygen-activated sludge plants responding to their questionnaire experienced *Nocardia* foaming. The average mixed liquor pH value of the air-activated sludge plants was 7; for the oxygen-activated sludge plants, the average pH value was 6.5. Lower mixed liquor pH values are typically encountered in oxygen-activated sludge plants because the mixed liquor is in contact with a gas phase high in CO_2. In addition, it is important to note that the aeration basins in oxygen-activated sludge systems always have subsurface withdrawal of mixed liquor, so that foam trapping occurs in them. The relative importance of foam trapping and mixed liquor pH on the high occurrence of *Nocardia* foaming in oxygen-activated sludge is not known precisely. However, an examination of Figures 105 and 111 suggests that foam trapping might well be the more important factor. Figure 105 suggests that the installation of trapping features in the laboratory pilot plant increased the *Nocardia* count approximately tenfold. Figure 111 suggests that a change in pH from about 7.0 (typical pH of air activated sludge systems) to a pH of 6.5 (typical pH of oxygen-activated sludge systems) would result in an increase in *Nocardia* count of about 20%, or about 1/50th of the effect on *Nocardia* count due to foam trapping. The occurrence of both low pH and foam trapping in the same system provides excellent conditions for *Nocardia* growth and retention.

It is observed that *Nocardia* foaming incidents often coincide with the onset of nitrification. This observation is explainable, based on both MCRT and pH effects. First, the onset of nitrification indicates that the aerobic MCRT is sufficiently high to allow the growth of nitrifying bacteria. Based on the data presented in Figure 110, this MCRT could also be high enough to allow

Table 33. Washout MCRT (or MCRT Used for *Nocardia* Control) of Various Activated Sludge Plants (Cha et al., 1992)

Source of Data	Temperature, °C	MCRT, days
Sacramento, CA – pilot plant	16	2.2
Sacramento, CA – pilot plant	24	1.6
Sacramento, CA – full scale (summer)	22	1.8
Sacramento, CA – full scale (winter)	18	2.2
San Francisco, CA – full scale	20	1.7

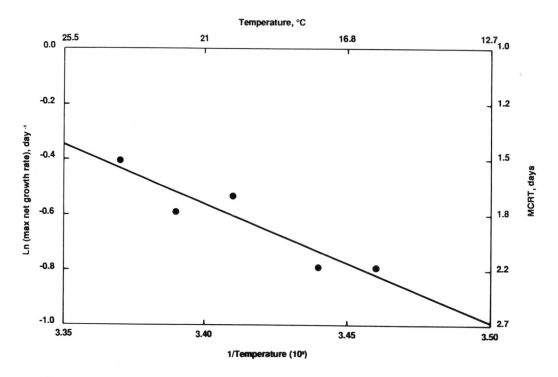

Figure 110. Arrhenius plot of *Nocardia* maximum net growth rate and temperature. (Cha et al., 1992). Reprinted by permission of the Water Environment Federation.

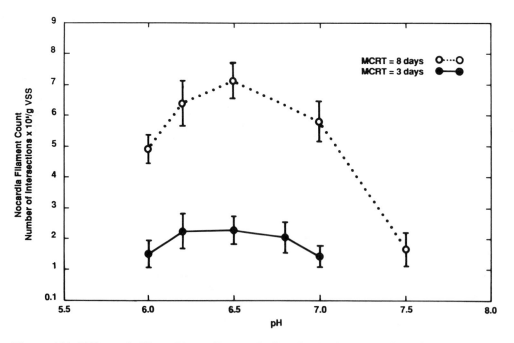

Figure 111. Effect of pH on *Nocardia* populations in the bench-scale activated sludge units. (Cha et al., 1992). Reprinted by permission of the Water Environment Federation.

Nocardia to grow. Secondly, the consumption of alkalinity by nitrification can result in depressed pH, which also encourages the growth of *Nocardia*.

The ability of the various types of selectors (aerobic, anoxic and anaerobic) to control *Nocardia* growth in activated sludge has been examined by a combination of pure culture experiments on *N. amarae* and laboratory, pilot, and full-scale experiments on activated sludge.

Blackall et al. (1991) grew *N. amarae* ASF3 and ASAC 1 in a chemostat with acetate as a sole carbon source over a wide range of dilution rates. The organisms were taken out of the chemostat and their batch acetate uptake rates determined under aerobic conditions (DO present), anoxic conditions (no DO present, but either NO_3^--N or NO_2^--N present) and anaerobic conditions (no DO, NO_3^--N, or NO_2^--N present). Batch substrate uptake rates suggest how well an organism will fare in a selector, where high substrate uptake rate is a key competitive advantage.

The aerobic batch acetate uptake rates for *N. amarae* are compared in Figure 112 to those of *Zoogloea ramigera*—a floc formerly found in selector activated sludge and shown by van Niekerk et al. (1987) to be responsible for the rapid uptake of soluble COD in an aerobic selector. The variation of batch acetate uptake rate with organism growth rate (dilution rate) is quite different for these two organisms. Of significance for *Nocardia* control purposes is the observation that only at very low dilution rates is the *N. amarae* ASF 3 unbalanced growth acetate uptake rate greater than that of *Z. ramigera*. It should be recognized that the dilution rates used in these studies (range 1 to 5 days) are significantly higher than the dilution rates applied in full-scale activated sludge systems (1/3 to 1/8 day^{-1}). Also, the uniform growth environment provided in a chemostat differs from the temporally and spatially variable environment provided in a full-scale system. This latter factor may result in differences in the physiological state of the organisms. Nevertheless, the data presented in Figure 112 suggest that an aerobic selector may be more effective in controlling *Nocardia* in activated sludge at a moderate MCRT than at high MCRT.

Cha et al. (1992) investigated the ability of an aerobic selector to control *Nocardia* populations in activated sludge at MCRTs of 5 and 10 days, using a four-compartment selector at a temperature of 20°C. A DO concentration of at least 5.0 mg/L was maintained in all selector compartments and in the main aeration basin. All units

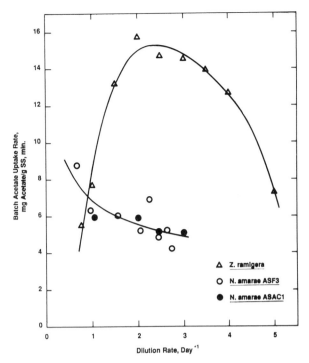

Figure 112. Comparison of *N. amarae* and *Z. ramigera* transient-state acetate uptake rates over a range of dilution rates. (Blackall et al., 1991). Reprinted by permission of the Water Environment Federation.

had *Nocardia* foam trapping features—i.e., subsurface aeration basin mixed liquor withdrawal and a baffle in front of the secondary clarifier weir. The aerobic selector was judged to be functioning properly as a selector by rapid soluble COD uptake rate in a batch test, control of SVI, and observance of typical amorphous zoogloeal colonies in the flocs.

Figure 113 shows that at a 5-day MCRT, the *Nocardia* count in the completely-mixed control unit increased from an initial value of 1×10^6 intersections/g VSS to 5×10^6 after about 2–3 MCRTs, and thereafter remained between 3.5×10^6 and 6×10^6 until the end of the experiment (after 7.5 MCRTs). The *Nocardia* count in the aerobic selector unit, on the other hand, declined gradually from its initial value of 1×10^6 to a value of 1×10^5 at the end of the experiment. The results indicate that, at a 5-day MCRT, the aerobic selector controlled *Nocardia*. When the aerobic selector was operated at a 10-day MCRT (Figure 114), effective *Nocardia* control was not observed. After one MCRT, *Nocardia* counts in the CSTR control system fluctuated between

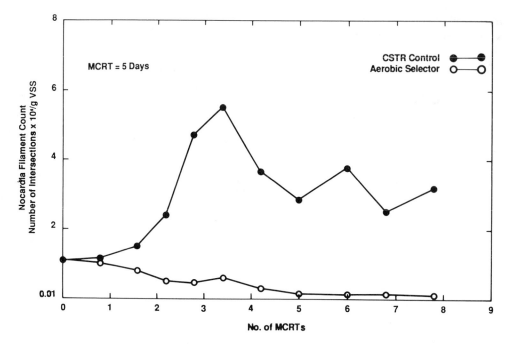

Figure 113. Effect of an aerobic selector on *Nocardia* populations at an MCRT of 5 days (bench-scale units). (Cha et al., 1992). Reprinted by permission of the Water Environment Federation.

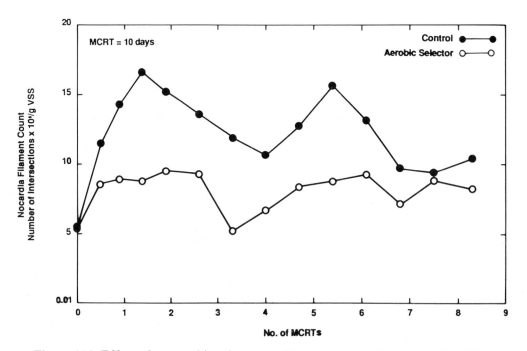

Figure 114. Effect of an aerobic selector on *Nocardia* populations at an MCRT of 10 days (bench-scale units). (Cha et al., 1992). Reprinted by permission of the Water Environment Federation.

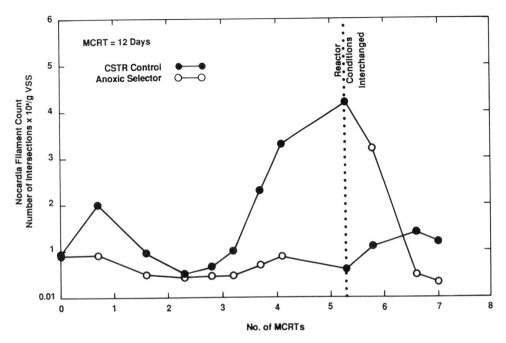

Figure 115. Effect of an anoxic selector on *Nocardia* populations in activated sludge in bench-scale units. (Cha et al., 1992). Reprinted by permission of the Water Environment Federation.

1.0×10^7 and 1.7×10^7, while those in the aerobic selector unit were somewhat lower (in the range of 5×10^6 to 1.0×10^7 intersections/g VSS) over the 9-MCRT long experiment. While the aerobic selector *Nocardia* populations were in a somewhat lower range than those in the control system, Figure 114 shows that toward the end of the experiment (between 7 to 9 MCRTs), the *Nocardia* counts in both units were very close to each other. On this basis, it can be stated that effective *Nocardia* control was not achieved by an aerobic selector at an MCRT of 10 days.

Operating results from the 27 MGD Upper Occoquan Sewage Authority (UOSA) Water Reclamation Plant (WRP) in Centreville, Virginia, are consistent with the results of Cha et al. (1992). The UOSA WRP utilizes aerobic selectors to control filamentous bulking [Daigger et al. (1985), Daigger and Nicholson (1990)]. The effectiveness of the selector in controlling the growth of filamentous organisms is demonstrated by SVIs in the full-scale system, which are routinely below 100 mL/g. Operating experience indicates that the selectors are effective in controlling *Nocardia* foaming in the summer (wastewater temperature generally greater than 20°C), when the MCRT is maintained within the 5 to 8 day range. However, *Nocardia* foaming occurs when the MCRT is greater than 10 days, even though the selector is being operated. *Nocardia* foaming does not generally occur in the winter (wastewater temperature 15°C or below), even though MCRTs of 10 to 15 days are maintained.

At MCRT values of ≥ 10 days, activated sludge systems operating at temperatures of > 20°C usually will nitrify. This offers the possibility of utilizing an anoxic selector for *Nocardia* control. This possibility was investigated by Cha et al. (1992) at an MCRT of 12 days. The anoxic selector design described earlier was used. It was shown to be functioning satisfactorily as a selector by soluble COD uptake and NO_3-N removal over the selector of approximately 40 mg/L and approximately 5 mg/L, respectively, control of SVI to < 100 mL/g, and presence of amorphous zoogloeal colonies in the activated sludge.

Figure 115 shows that the CSTR control unit exhibited *Nocardia* counts that ranged between 1×10^6 and 4×10^6 intersections/g VSS over a period of 7 MCRTs (generally increasing throughout the experiment). In the same period, the anoxic selector system *Nocardia* count remained between 5×10^5 and 1×10^6 intersections/g VSS. After 66 days (5.5 MCRTs), the contents of the reactors were interchanged so that the CSTR control activated sludge was placed in the anoxic selector system and vice versa. The *Nocardia* count in the activated sludge from the control unit immediately and rapidly decreased under anoxic selector conditions; the anoxic selector sludge, placed in the CSTR control unit, showed a gradual increase in *Nocardia* count to levels above those observed during the time that this sludge had been in the anoxic selector unit. These experiments suggest that properly designed anoxic selectors may be effective in controlling *Nocardia* in activated sludge. They are also consistent with the pure culture data of Blackall et al. (1991) on *N. amarae*; these workers showed that *N. amarae* could not take up acetate in anoxic batch tests (Figure 116) and that it denitrified at rates that were 2 to 3 orders of magnitude less than *Z. ramigera* (van Niekerk et al., 1987), and only incompletely to NO_2-N, rather than N_2 gas (Table 28).

Sezgin and Karr (1986) converted the first pass of the four-pass aeration basin in a nitrifying activated sludge plant at Utoy Creek, Georgia, to an

anoxic zone by reducing the air flow to the coarse bubble diffusers. On three occasions out of five, Richards et al. (1990) reported that this was successful in reducing foam coverage of the aeration basins. A truly anoxic zone was not obtained (0.2 to 0.5 mg DO/L was present) because of the use of aeration for mixing; also, low aeration rates in the "anoxic zone" limited the extent of nitrification in the plant, meaning that less NO_3-N was recycled back to the anoxic zone in the RAS.

Results from other full-scale plants utilizing anoxic selectors indicate the dual impact of the selector and of foam trapping/recycle. The 17 MGD Rock Creek Advanced Wastewater Treatment Plant (AWT) uses an anoxic selector for filament control and alkalinity recovery [Daigger and Nicholson (1990)]. An accumulation of *Nocardia* foam is observed at this plant during periods when secondary scum is recycled to the plant headworks. Similarly, *Nocardia* foaming was observed during pilot studies for the 40-MGD VIP plant in Norfolk, Virginia [Daigger et al. (1988)], and has also been observed in the recently completed full-scale facility. Both anoxic and anaerobic selectors are incorporated into the VIP pilot and full-scale facilities. Foam trapping occurred in both the pilot and full-scale units, and may explain the *Nocardia* foaming incidents. *Nocardia* foaming at other full-scale nutrient removal plants (which incorporate anoxic zones for nitrogen removal) is often attributed to foam trapping caused by the placement of baffles. Taken together, these results indicate that anoxic selectors may help to control *Nocardia* growth, but that their positive effects may be offset by foam trapping and recycle.

The data on the effectiveness of anaerobic selectors are not as clear-cut as for aerobic and anoxic selectors. Blackall et al. (1991) showed that *N. amarae* neither grew nor took up acetate under anaerobic conditions (Figure 116). However, in both laboratory-scale and full-scale activated sludge experiments (Pitt and Jenkins, 1990), only small and inconsistent *Nocardia* reductions were obtained using an anaerobic selector.

Figure 117 shows laboratory-activated sludge results from a series of anaerobic selector tests. In one out of four of these runs, *Nocardia* counts were reduced in the anaerobic selector unit sooner and to a lower level than in the control. These experiments were plagued by the growth of *Thiothrix* spp. on the septic influent sewage. It is postulated that the *Thiothrix* spp. competed successfully with *Nocardia* spp. for the soluble sub-

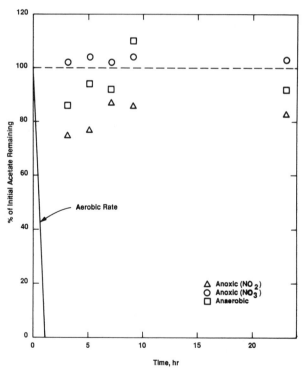

Figure 116. Transient-state acetate uptake of *N. amarae* ASF3 under aerobic, anoxic, and anaerobic conditions. (Blackall et al., 1991). Reprinted by permission of the Water Environment Federation.

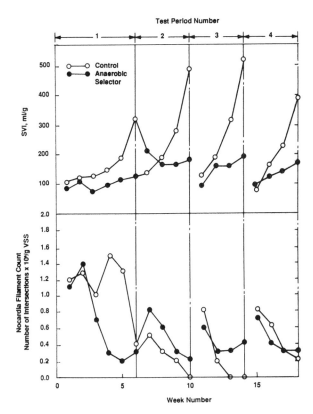

Figure 117. SVI and *Nocardia* counts for bench-scale anaerobic selector experiments. (Pitt and Jenkins, 1990). Reprinted by permission of the Water Environment Federation.

strate. Figure 117 shows that *Nocardia* counts decreased in both the anaerobic selector unit and the control system as the SVI increased due to *Thiothrix* spp. growth.

Full-scale anaerobic selector experiments were conducted at the San Francisco South East Water Pollution Control Plant during the periods May through Oct. 1987, and May through Oct. 1988. This 6 equal-size-compartment aeration basin oxygen-activated sludge plant has 8 treatment trains, 4 of which were converted to the A/O process by making the first compartment anaerobic (replacing the aerator with a mixer and feeding the O_2 gas to the second compartment).

During 1987, three anaerobic selector operating conditions were examined (Table 34). These ranged from an initial selector F/M and hydraulic detention time of 6.6 to 13.2 kg BOD_5/kg MLVSS,day and 13 to 25 min, respectively. While it was possible to separate the plant into two halves, the results were confounded by cross-contamination between the control and the anaerobic selector system by:

- Control RAS spilling over a common wall into anaerobic selector RAS and
- Foam spilling from the control mixed liquor into the anaerobic selector mixed liquor over a dividing gate in the mixed liquor channel. The RAS cross-contamination was rectified during the 1987 testing.

Table 35 shows that *Nocardia* levels, as assessed by two counting methods and secondary clarifier foam coverage, were substantially lower in the anaerobic selector system than in the control during Test Period 1, which had the highest selector F/M and the shortest hydraulic detention time. This was also the test period with the greatest differences between anaerobic selector foam height in the foam test, secondary effluent SS concentration, and anaerobic selector soluble orthophosphate release. In none of the test periods did the control and anaerobic selector SVI differ significantly. At no time did the control and anaerobic selector systems show any difference in phosphate removal.

The full-scale anaerobic selector test was repeated in the summer of 1988 using the test conditions shown in Table 36. Cross-contamination due to foam spilling from the control system into the anaerobic selector system occurred for the first 5 weeks of this experiment.

The 1988 results (Table 37) show that the anaerobic selector again reduced *Nocardia* levels compared to the control. Again, the selector functioned best at the highest F/M and lowest hydraulic detention time investigated (2.1 kg BOD_5/kg MLVSS,day and 15 min, respectively).

Even at the best conditions tested, *Nocardia* was still the dominant filamentous organism in the mixed liquor. In spite of this, significant reductions in secondary clarifier foam coverage, mixed liquor *Nocardia* count, numerical scale rating, and foam height in a foam test were observed for the best conditions. Accompanying these decreased anaerobic selector system *Nocardia* levels was significant decrease in secondary effluent SS, likely because of less foam escaping into the secondary effluent.

Mamais and Jenkins (1992) have determined the minimum MCRT that activated sludge will exhibit enhanced biological phosphorus removal (EBPR) over a range of temperatures. The occurrence of EBPR is necessary for the proper functioning of an anaerobic selector, because EBPR is characteristic of the growth of organisms capable of taking up biodegradable soluble organics in the anaerobic zone. Without such uptake, the

Table 34. Full-Scale Anaerobic Selector Experiment Test Conditions, San Francisco SEWPCP, 1987 (Pitt and Jenkins, 1990)

Test Period	Test Duration, weeks	Number of Aeration Trains Utilized: Total	Anaerobic Selector	Control	MCRT, days	Overall F/M, kg BOD$_5$/kg MLVSS, day	Selector F/M, kg BOD$_5$/kg MLVSS, day	Selector Average Hydraulic Detention Time, min
1	11	4	2	2	1.4	2.2	13.2	13
2	5	6	4	2	2.0	1.3	7.8	19
3	6	8	4	4	2.3	1.1	6.6	25

Table 35. Results of Full-Scale Anaerobic Selector Experiment, San Francisco SEWPCP, 1987 (Pitt and Jenkins, 1990)

Test Period	Average Nocardia Count, Number of Intersections × 10^4/g VSS — Anaerobic Selector	Control	Average Secondary Clarifier Foam Coverage, % — Anaerobic Selector	Control	Average Nocardia Numerical Scale Rating — Anaerobic Selector	Control	Average Foam Test Height, mL — Anaerobic Selector	Control	Average Anaerobic Selector (First Stage) Soluble Orthophosphate Release, mg P/L — Anaerobic Selector	Control	Average Secondary Effluent SS, mg/L — Anaerobic Selector	Control	Average SVI, mL/g — Anaerobic Selector	Control
1	3.7	5.3	5	28	3.0	4.5	530	650	3.4	0	19	28	126	132
2	4.3	5.0	10	17	3.0	3.6	570	620	2.3	0	21	25	111	121
3	4.8	5.3	27	36	3.5	4.0	600	610	0.8	0	19	29	108	118

Table 36. Full-Scale Anaerobic Selector Experiment Test Conditions, San Francisco SEWPCP, 1988 (Pitt and Jenkins, 1990)

Test Period	Test Duration, weeks	Number of Aeration Trains Utilized: Total	Anaerobic Selector	Control	MCRT, days	Overall F/M, kg BOD$_5$/kg MLVSS, day	Selector F/M, kg BOD$_5$/kg MLVSS, day	Selector Average Hydraulic Detention Time, min
1	13	6	3	3	1.6	1.6	10	22
2	5	4	2	2	1.4	2.1	13	15
3	4	6	3	3	1.7	1.5	9	21

Table 37. Results of Full-Scale Anaerobic Selector Experiment, San Francisco SEWPCP, 1987 (Pitt and Jenkins, 1990)

Test Period	Average Nocardia Count, Number of Intersections × 10^4/g VSS Anaerobic Selector	Control	Average Secondary Clarifier Foam Coverage, % Anaerobic Selector	Control	Average Nocardia Numerical Scale Rating Anaerobic Selector	Control	Average Foam Test Height, mL Anaerobic Selector	Control	Average Anaerobic Selector (First Stage) Soluble Orthophosphate Release, mg P/L Anaerobic Selector	Control	Average Secondary Effluent SS, mg/L Anaerobic Selector	Control	Average SVI, mL/g Anaerobic Selector	Control
1	4.7	5.5	10	15	3.2	4.0	560	580	2.2	0	17	17	113	128
2	4.9	6.2	0	7	3.0	4.4	580	620	2.9	0	24	34	117	132
3	6.7	7.8	0	9	3.7	4.5	570	600	2.1	0	20	46	110	122

beneficial effects of the selector will not be exhibited. Figure 118 presents an Arrhenius plot of the washout MCRT of EBPR. These data can be used to establish the minimum MCRT at which an anaerobic selector can be operated since, as just discussed, EBPR is necessary to take up the soluble substrate in an anaerobic selector. Figure 119 shows that the washout MCRT for *Nocardia*

spp. in activated sludge at a range of temperatures (Cha et al., 1992) is very close to the washout MCRT for EBPR. Thus, a plant being operated at a low MCRT for *Nocardia* control will very likely be unable to consistently support EBPR, and therefore would not provide satisfactory conditions for an anaerobic selector.

The anaerobic selector experiments of Pitt and

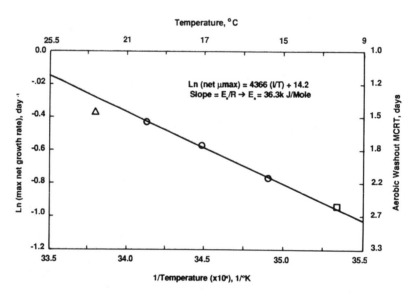

Figure 118. Arrhenius plot of washout MCRT for enhanced biological phosphate removal in activated sludge. (Mamais and Jenkins, 1992).

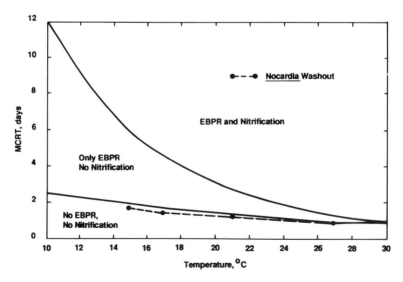

Figure 119. Limiting MCRT values for nitrification, EBPR and *Nocardia* growth at a range of temperatures.

Jenkins (1990) at San Francisco SEWPCP were carried out at MCRT values in the range 1.4 to 2.3 days (Tables 34 and 36) — values that had previously been established at this plant to provide *Nocardia* control by washout. It is likely that these MCRTs were too low to provide for the establishment of a consistently performing anaerobic selector (Figure 119).

Reference has previously been made to the analogy that can be drawn between *Nocardia* foaming (flotation) in activated sludge and the use of hydrophobic collectors for the separation of minerals (beneficiation) from ores by selective flotation (classification). This principle has been used to remove *Nocardia* from activated sludge. Pretorius (1987) suggested that the buoyancy of *Nocardia* could be used in a selector based on the principle of a flotation unit to selectively separate and remove *Nocardia* from settleable activated sludge flocs. He suggested that if the float from such a "classifying selector" were not returned to the aeration basin, a selection against *Nocardia* would be obtained. Pretorius and Laubscher (1987) investigated this principle in 200 L mechanically-aerated activated sludge plants, one of which was equipped with a 16.7 L fine bubble flotation cell between the aeration basin and the secondary clarifier. The foam level in the classifying selector system was reduced to zero after about 2 days, while the control system continued to foam severely. While the classifying selector had no long-term (24 days) effect on the MLSS level, there was a decrease (from about 4500 mg/L to 4000 mg/L) during the first few days following the installation of the classifying selector. This observation suggests that a classifying selector applied to a long-standing, severe *Nocardia* foaming problem with much accumulated foam may initially waste out more than the desirable amount of solids inventory. In such circumstances it may well be good practice to vacuum off as much accumulated foam as possible, then start using the classifying selector.

Selective foam wasting was employed successfully by Richards et al. (1990) at the Utoy Creek, GA activated sludge plant. This plant was operated in a sludge reaeration/step-feed mode (Figure 120) with RAS only in the first pass of the aeration basin. Foam was allowed to pass out of the aeration basin into an unused basin, where it was allowed to concentrate. Foam SS concentration was between 2% and 3%, and the foam was found to dewater well with the centrifuges normally used for WAS. Using the classifying selector to control foam allowed the plant to increase

aeration rates and MCRT values (from 6 to 20 days), so that complete nitrification was possible for the first time (Figure 121). It was observed that once the MCRT was raised to greater than 20 days, the severe foaming ceased and no foam wasting occurred. From a previous article written on this plant (Sezgin and Karr, 1986), it is suspected that an industrial waste containing a slowly degradable surfactant was present which was only degraded completely at high MCRT. The *Nocardia* foaming may have been stabilized by the presence of the surfactant at the lower MCRT.

The technique of RAS chlorination is not very successful in controlling *Nocardia* in activated sludge because the *Nocardia* filaments are largely contained inside the activated sludge flocs. Thus, they are inaccessible to the chlorine unless a high enough dose (actually an overdose) is applied to break up the flocs. Such doses are not recommended because they degrade effluent quality and diminish treatment capacity. Some benefit from RAS chlorination in controlling *Nocardia* can be achieved where foam trapping exists, because this generates free-floating *Nocardia* filaments that are accessible to the chlorine.

A far more effective use of chlorine in *Nocardia* control is to apply the chlorine as a fine spray directly to the aeration basin surface. This approach has been used successfully at several plants in the USA including the 23rd Avenue plant, Phoenix AZ, Stamford CT, Ocean County NJ, and Trenton NJ (Albertson, 1991). At the 23rd Avenue plant in Phoenix, AZ, spray hoods (Figure 122) were placed across the end of the 3rd pass of the 4-pass aeration basin. The foam entered the hood with the mixed liquor flow and was sprayed at a rate of approximately 10 L/min with a chlorine solution (from a gas chlorinator) containing 2000–3000 mg Cl_2/L. Using a chlorine dose in the range 0.5 to 1.0 mg Cl_2/L based on the wastewater flow, *Nocardia* foam was eliminated on several occasions within 1–2 days without any effluent quality deterioration or loss of treatment efficiency of any kind. Albertson (1991) suggests that the chlorine surface spray technique for *Nocardia* control may only be effective when *Nocardia* trapping and foam recycle features are eliminated from the system. It may be possible to enhance the effectiveness of the chlorine surface spray by installing fine bubble aeration to enhance the *Nocardia* foaming just prior to or within the chlorine surface spray system.

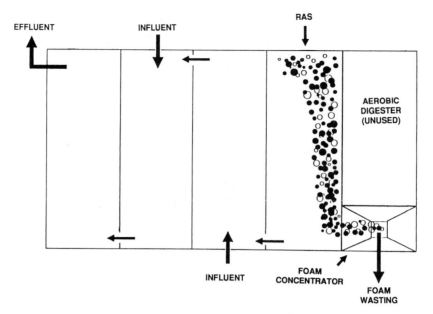

Figure 120. Plan view of Utoy Creek, GA aeration basins with selective foam wasting. (Richards et al., 1990). Reprinted by permission of the Water Environment Federation.

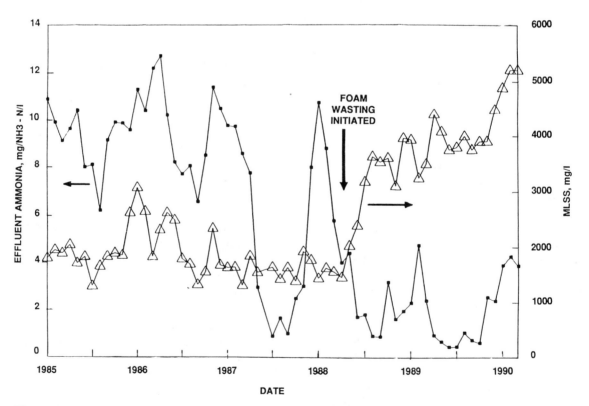

Figure 121. Effluent ammonia and mixed liquor SS concentrations over time at Utoy Creek, GA before and after selective foam wasting. (Richards et al., 1990). Reprinted by permission of the Water Environment Federation.

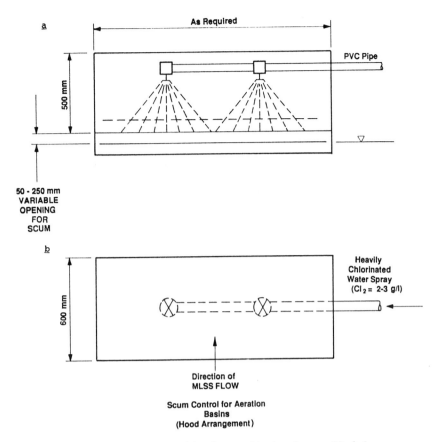

Figure 122. *Nocardia* control by foam chlorination at 23rd Ave Wastewater Treatment Plant, Phoenix, AZ: *a.* side view; *b.* top view.

SUMMARY

In many ways our current knowledge of the causes and controls of *Nocardia* growth and foaming in activated sludge is in a state similar to our knowledge of the causes and controls of filamentous bulking at the time the first edition of this manual was written. Some of the causes and control methods are fairly well understood. Some of the control methods have been tried out both in pilot plants and in prototype plants; others still await full-scale application. Because of this state of affairs, we think it is useful to present a summary that both presents our current knowledge as well as identifies areas where further work is needed.

Generally, it can be stated that far more is known about how to get rid of *Nocardia* spp. than what causes it to grow. Laboratory studies have confirmed that *Nocardia amarae* can grow well on simple soluble substrates such as acetate. Anecdotal reports from many wastewater treatment plants suggest that growth occurs also on lipophilic substrates (i.e., "oil and grease"). We need further data in this area.

The importance of physical plant details and foam disposal practices on *Nocardia* populations and foaming in activated sludge (especially those that allow for foam trapping and recycle) has been firmly established. Intentional foaming and foam removal has been used successfully for *Nocardia* control in a limited number of prototype plants. Additional applications are needed to test the range of this method. Foam destruction by surface chlorination is also very limited in its prototype application. Application rates, points of application, and effectiveness in systems that trap *Nocardia* foam need to be assessed on a full scale. *Nocardia* control by reducing MCRT has long been employed in practice. Research verified that this is a sound technique, and indicated that the washout MCRT is a function of aeration basin temperature. In general, very low MCRT values (below 3 days) are required for this control method to be effective and, in the presence of foam trapping and recycle, it is virtually impossible to completely wash out the *Nocardia*.

Aerobic selectors have a limited MCRT (fairly low) range of application for *Nocardia* control and have only been tried out in the laboratory. Full-scale results are needed. Anoxic selectors have worked well in controlling *Nocardia* both in laboratory and a limited number of full-scale applications. Anaerobic selectors have not yet been successfully applied for *Nocardia* control either in the laboratory or in full-scale plants, although the observation that *N. amarae* does not take up acetate under anaerobic conditions is an indication that they should be effective. Previous full-scale anaerobic selector experiments were most likely done at too low an MCRT for establishment of the polyphosphate accumulation mechanism necessary for proper anaerobic selector functioning. The ability of any kind of selector to control *Nocardia* foaming in a full-scale plant with extensive foam trapping and recycle has not yet been assessed.

The fate of *Nocardia* in waste activated sludge treated in anaerobic digesters has not been determined.

While this chapter has presented a conceptual outline of the mechanism of how *Nocardia* stabilizes foams in activated sludge, more information is needed on the precise mechanisms involved. This could lead to the development of successful foam control techniques for both activated sludge and anaerobic digestion systems.

Hopefully we will be able to fill most of the "needs" expressed above in the writing of the third edition of this manual!!

Bibliography and References

This section, besides containing references cited in the text, also contains a more complete coverage of the literature on activated sludge bulking, filamentous organism growth and *Nocardia* growth and foaming.

Adamse, A.P. (1968) "Bulking of Dairy Waste Activated Sludge," *Water Res. 2*, 715.

Albagnac, G. and Morfaux, J.N. (1980) "Traitabilitié Comparée en Aeration Prolongée et en Contact-Stablisation des Eaux Résiduaires de Brasserie," *Trib. Cebedeau, 33*, 63.

Albertson, O. E. (1987) "The Control of Bulking Sludges: From the Early Innovators to Current Practice," *J. Water Polln. Control Fedn., 59*, 172.

Albertson, O. E. (1990) "Bulking Sludge Control—Progress, Practice and Problems," *Water Sci. and Technol., 23* (Kyoto), 835.

Albertson, O.E. (1991) Privately communicated, Salt Lake City, UT.

Albertson, O.E. and Hendricks, P. (1992) "Bulking and Foaming Organism Control at Phoenix, AZ WWTP." Presented at 16th Intl Assoc. for Water Polln Res. and Control, Washington DC, USA, May 1992.

Al-Diwamy, L. J. and Cross, T. (1978) "Ecological Studies of *Nocardia* Foams and Other Actinomycetes in Aquatic Habitats," in *Nocardia and Streptomyces*; M. Modarski, W. Kurytowicz, and J. Jeljaszewicz (Eds.), Gustav Fischer Verlag, Stuttgart, Germany, 153.

Allen, L.A. (1944) "The Bacteriology of Activated Sludge," *J. Hyg., 43*, 424.

Anon. (1969) "Milwaukee Mystery: Unusual Operating Problem Develops," *Water and Sewage Works, 116*, 213.

Anon., FMC Corporation (1973) "Bulking Control with Hydrogen Peroxide: Case History, Water Pollution Control Plant, City of Petaluma, Sonoma, CA," Tech. Data Polln. Control Release No. 41.

Anon. (1974) *Bergey's Manual of Determinative Bacteriology*, 8th Ed., R.E. Buchanan and N. E. Gibbons, Eds., The Williams and Wilkins Co., Baltimore, MD.

Anon. FMC Corporation (1976) "Sludge Bulking Cure: Hydrogen Peroxide," Tech. Data Polln. Control Release No. 95.

Anon. (1979) "Hydrogen Peroxide Solves Bulking Problem at Coors Waste Treatment Plant," *Food Engineering*, Nov 1979 and FMC Corporation Tech. Data Polln. Control Release No. 117.

Anon. (1981) *Manual of Methods for General Bacteriology*, Amer. Soc. for Microbiol., Washington, DC.

Anon. (1983) "Thames Water Uses Chlorine to Control Bulking Sludge," *Water Res. News, 10*, 6, Water Res. Centre, Medmenham, Bucks, England.

Anon. (1985) *"Clarifier Design"* Manual of Practice FD-8. Water Polln. Control Fedn., Arlington, VA.

Anon. (1990) *"Wastewater Biology: The Microlife,"* Water Polln. Control Fedn., Alexandria, VA.

Ardern, E. and Lockett, W.T. (1914a) "Experiments on the Oxidation of Sewage Without the Aid of Filters," *J. Soc. Chem. Ind., 33*, 523.

Ardern, E. and Lockett W.T. (1914b) "The Oxidation of Sewage Without the Aid of Filters," *J. Soc. Chem. Ind., 33*, 112.

Argaman, Y. and Eckenfelder, W.W. Jr. (1989) "The Effect of Selector Removal on Biomass Composition," *Res. J. Water Polln Control Fedn., 61*, 1731.

ATV Working Group 2.6.1. Report, "Prevention and Control of Bulking Sludge and Scum," *Korrespondenz Abwasser, 36* Jahrgang, 165.

Baines, S., Hawkes, H.A., Hewitt, C.H. and Jenkins, S.H. (1953) "Protozoa as Indicators in Activated Sludge Treatment," *Sewage Ind. Wastes, 25*, 1023.

Banoub, A. (1982) "Reducing Energy Consumption: How 2 Communities Did It; Woonsocket, R.I.," Deeds and Data; *Water Polln. Control Fedn., Highlights, 4*, 11.

Barnard, J.L. (1978) "Solving Sludge Bulking Problems," *Water Polln. Control, 77*, 103.

Barnes, D. and Goronszy, M.C. (1980) "Continuous Intermittent Wastewater Systems for Municipal and Industrial Effluents," *Publ. Hlth. Engr, 8,* 20.

Becker, J.G. and Shaw, C.G. (1955) "Fungi in Domestic Sewage Treatment Plants," *Appl. Microbiol., 3,* 173.

Beebe, R.D. and Jenkins, D. (1981) "Control of Filamentous Bulking at the San Jose/Santa Clara Water Pollution Control Plant," presented at 53rd Ann. Conf., Calif. Water Polln. Control Assoc., Long Beach, CA.

Beebe, R.D., Jenkins, D. and Daigger, G.T. (1982) "Activated Sludge Bulking Control at the San Jose/Santa Clara, Calif. Water Polln. Control Plant," presented at 55th Ann. Conf. Water Polln. Control Fedn., St. Louis, MO.

Beebe, R.D. (1983) Privately communicated, San Jose/Santa Clara Water Polln. Control Plant, San Jose, CA.

Benefield, L.D., Randall, C.W. and King, P.H. (1975) "The Stimulation of Filamentous Micro-organisms in Activated Sludge by High Oxygen Concentrations," *Water Air and Soil Polln., 5,* 113.

Blackall, L.L. (1986) "Actinomycete Scum Problems in Activated Sludge Plants," Ph.D. Dissertation, University of Queensland, Australia.

Blackall, L.L., Parlett, J.H., Hayward, A.C., Minnikin, D.E., Greenfield, P.F. and Harbers, A.E. (1989) "*Nocardia pinensis* sp. nov., an Actinomycete found in Activated Sludge Foams in Australia," *J. Gen. Microbiol., 135,* 1547.

Blackall, L.L., Tandoi, V. and Jenkins, D. (1991) "Continuous Culture Studies With *Nocardia amarae* From Activated Sludge and their Implications for Foaming Control," *Res. J. Water Polln. Control Fedn., 63,* 44.

Blackbeard, J.R. and Ekama, G.A. (1984) "Preliminary Report on Filamentous Micro-organisms Responsible for Bulking and Foaming in Activated Sludge Plants in Southern Africa," Dept. of Civil Eng. Univ. of Cape Town, RSA.

Blackbeard, J.R., Ekama, G.A. and Marais, G.v.R. (1986) " A Survey of Bulking and Foaming in Activated Sludge Plants in South Africa," *Water Polln. Control, 85,* 90.

Blackbeard, J.R., Ekama, G.A. and Marais, G.v.R. (1986) "Identification of Filamentous Organisms in Nutrient Removal Activated Sludge Plants in South Africa," *Water SA, 14,* 29.

Bode, H. (1983) "The Use of Chlorine for Bulking Control." Presented at the the Institut für Siedlungswasserwirtshaft, Univ. of Hannover, Germany, 26 Jan 1983.

Boyle, W.C. (1983), Discussion of Lee, S-E, Koopman, B.L., Jenkins, D. and Lewis, R.F. (1983), "The Interrelated Effects of Aeration Basin Configuration on Activated Sludge Bulking at Low Organic Loading," *Water Sci. Technol. (Capetown)., 14,* 953.

Buhr H.O., (1982) "Solids Distribution in a Step Feed Plant," Internal Rept, Greeley and Hansen, Phoenix, AZ.

Burke, R.A., Dold, P.L. and Marais, G.v.R. (1986) "Biological Excess Phosphorus Removal in Short Sludge Age Activated Sludge Systems," Res. Rept No. W58, Dept. of Civil Eng, Univ. of Capetown, RSA.

Butterfield, C.T. (1935) Studies of Sewage Purification: II "*A Zoogloea-Forming Bacterium Isolated from Activated Sludge,*" U.S. Pub. Hlth Rpt, No. 50, 671.

Butterfield, C.T., Ruchoft, C.C. and McNamee, P.D. (1937) "*Studies of Sewage Purification*: *VI. Biochemical Oxidation by Sludges Developed by Pure Cultures of Bacteria Isolated from Activated Sludge,*" U.S. Pub. Hlth Rpt. No. 52, 387.

Cairns, W.L., Cooper, D.G., Zajic, J.E., Wood, J.M. and Losaric, N. (1982) "Characterization of *Nocardia amarae* as a Biological Coalescing Agent of Oil-water Emulsions," *Appl. Environ. Microbiol., 43,* 362.

Calaway, W.T. (1963) "Nematodes in Wastewater Treatment," *J. Water Polln. Control Fedn. 35,* 1006.

Calaway, W.T. (1968) "The Metazoa of Waste Treatment Processes—Rotifers," *J. Water Polln. Control Fedn. 40,* 412.

Campbell, H.J. Jr., Troe, D., Gray, R., Jenkins, D. and Kirby, C.W. (1985) "InBasin Chlorination for Control of Activated Sludge Bulking in Industrial Waste Treatment Plants," Proc. 40th Ind. Waste Conf., Purdue Univ. West Lafayette, IN, Butterworth, 759.

Caropreso, F.E., Raleigh, C.W. Brown, J.C. (1974) "Attack Bulking Sludge with H_2O_2 and a Microscope," *Bull. Calif. Water Polln. Control Assoc.,* 44.

Carter, J.L. and McKinney, R.E. (1973) "Effects of Iron on Activated Sludge Treatment," *J. Environ. Eng. Div. Amer. Soc. Civil Eng., 99,* E2, Proc. Paper 96–79, 135.

Cha, D.K., Jenkins, D., Lewis, W.P. and Kido, W.H. (1992) "Process Control Factors Influencing *Nocardia* Populations in Activated Sludge" *Water Env. Res., 64*, 37.

Chambers, B. (1982) "Effect of Longitudinal Mixing and Anoxic Zones on Settleability of Activated Sludge," Chapter 10 in *Bulking of Activated Sludge: Preventative and Remedial Methods*, Eds, B. Chambers and E.J. Tomlinson, Ellis Horwood Ltd., Chichester, England.

Chiesa, S.C. and Irvine, R.L. (1985) "Growth and Control of Filamentous Microbes in Activated Sludge: an Integrated Hypothesis," *Water Res., 19*, 471.

Chudoba, J., Ottova, V. and Madera, V. (1973a) "Control of Activated Sludge Filamentous Bulking. I. Effect of Hydraulic Regime or Degree of Mixing in an Aeration Tank." *Water Res., 7*, 1163.

Chudoba, J., Grau, P. and Ottova, V. (1973b) "Control of Activated Sludge Filamentous Bulking. II. Selection of Micro-organisms by Means of a Selector," *Water Res., 7*, 1389.

Chudoba, J., Blaha, J. and Madera, V. (1974) "Control of Activated Sludge Filamentous Bulking. III. Effect of Sludge Loading," *Water Res., 8*, 231.

Chudoba, J., Dohanyos, M. and Grau, P. (1982) "Control of Activated Sludge Filamentous Bulking. IV. Effect of Sludge Regeneration," *Water Sci. Technol., 14*, 73.

Chudoba, J. (1985) "Control of Activated Sludge Filamentous Bulking—VI. Formulation of Basic Principles," *Water Res., 19*, 1017.

Chudoba, J. and Wanner, J. (1987) "Discussion of Albertson, O.E. (1987), The Control of Bulking Sludge: from the Early Innovators to Current Practice," *J. Water Polln. Control Fedn, 59*, 172.

Chudoba, J. (1989) Activated Sludge Bulking Control, Chapter 6 in *Encyclopedia of Environmental Control Technology*, Vol.3 *Wastewater Treatment Technology*, Ed., P.N. Chereminisoff, Gulf Publishing Co, Houston, TX.

Clesceri, L.S. (1963) "Effect of Fermentation Variables on Properties of Activated Sludge," Ph.D. Dissertation, Univ. of Wisconsin, Madison, WI.

Cole, C.A., Stamberg, J.B. and Bishop, D.F. (1973) "Hydrogen Peroxide Cures Filamentous Growth in Activated Sludge," *J. Water Polln. Control Fedn., 45*, 829.

Cooke, W.B. and Ludzack, F.J. (1958) "Predaceous Fungi Behavior in Activated Sludge Systems," *Sewage Ind. Wastes, 30*, 1490.

Cooke, W.B., and Pipes, W.O. (1970) "The Occurrence of Fungi in Activated Sludge," *Mycopath. Mycop. Appl., 40*, 249.

Cooper, P.F., Collinson, B. and Green, M.K. (1977) "Recent Advances in Sewage Effluent Denitrification: Part II," *Water Polln. Control, 76*, 389.

Costerton, J.W. and Irvin, T.R. (1981) "The Bacterial Glycolax in Nature and Disease," *Ann. Rev. Microbiol., 32*, 299.

Curds, C.R., Cockburn, A. and Vandyke, J.M. (1968) "An Experimental Study of the Role of the Ciliated Protozoa in the Activated Sludge Process," *Water Polln. Control, 67*, 312.

Curds, C.R. (1969) "*An Illustrated Key to the British Freshwater Ciliated Protozoa Commonly Found in Activated Sludge*," Tech. Paper No.12, Water Polln Res., Ministry of Technology, London.

Curds, C.R., and Fey, G.J. (1969) "The Effect of Ciliated Protozoa on the Fate of *Escherichia coli* in the Activated Sludge Process," *Water Res., 3*, 853.

Curds, C.R., and Cockburn, A. (1970) "Protozoa in Biological Sewage Treatment Processes—I. A Survey of the Protozoa Fauna of British Percolating Filters and Activated Sludge Plants," *Water Res., 4*, 225.

Curds, C.R., and A. Cockburn (1970a) "Protozoa in Biological Sewage Treatment Processes—II. Protozoa as Indicators in the Activated Sludge Process," *Water Res., 4*, 237.

Curds, C.R. (1973) "The Role of Protozoa in the Activated Sludge Process," *Amer. Zool., 13*, 161.

Curds, C.R. (1975), Chapter 5 "Protozoa" in "*Ecological Aspects of Used-Water Treatment.*" Vol.I. "The Organisms and Their Ecology," Eds, Curds, C.R. and Hawkes, H.A., Academic Press, New York.

Curtis, E.J.C, (1969) "Sewage Fungus: Its Nature and Effects," *Water Res., 3*, 280.

Cyrus, Z. and Sladka, A. (1970) "Several Interesting Organisms Present in Activated Sludge," *Hydrobiologia, 35*, 383.

Daigger, G.T., Robbins, M.H. Jr. and Marshall, B.R. (1985) "The Design of a Selector to Control Low F/M Filamentous Bulking," *J. Water Polln. Control Fedn., 57*, 220.

Daigger, G.T. and Roper, R.E. Jr. (1985) "The Relationship Between SVI and Activated Sludge Settling Characteristics," *J. Water Polln Control Fedn, 57*, 859.

Daigger, G.T., Waltrip, G.D., Romm, E.D., and

Morales, L.A. (1988) "Enhanced Secondary Treatment Incorporating Biological Nutrient Removal" *J. Water Polln Control Fedn., 60,* 1833.

Daigger, G.T. and Nicholson, G.A. (1990) "Performance of Four Full-Scale Nitrifying Wastewater Treatment Plants Incorporating Selectors," *Res. J. Water Polln. Control Fedn., 62,* 676.

Deekyne, C.H.W., Patal, M.A. and Krichten, D.J. (1983) *"Demonstration of Biological Phosphorus Removal by the A/O Process at 70 MGD Patapsco Wastewater Treatment Plant,"* Whitman, Requardt and Associates, Baltimore, MD.

De L.G. Solbe, J.F. (1975) Chapter 8 "Annelida" in *"Ecological Aspects of Used-Water Treatment".* Vol.I. "The Organisms and Their Ecology," Eds, Curds, C.R. and Hawkes, H.A., Academic Press, New York.

Dhaliwal, B.S. (1979) *"Nocardia amarae* and Activated Sludge Foaming," *J. Water Polln. Control Fedn., 51,* 344.

Dias, F.F., and Bhat, J.V. (1964) "Microbial Ecology of Activated Sludge. I. Dominant Bacteria," *Appl. Microbiol., 12,* 412.

Dias, F.F., Dondero, N.C. and Finstein, M.S. (1968) "Attached Growth of *Sphaerotilus* and Mixed Populations in a Continuous-Flow Apparatus," *Appl. Microbiol., 16,* 1191.

Dick, R.I. and Vesilind P.A. (1969) "The Sludge Volume Index—What is it?," *J. Water Polln. Control Fedn., 41,* 1285.

Dold, P.L. and Marais, G.v.R. (1986) "Evaluation of the General Activated Sludge Model Proposed by the IAWPRC Task Group," *Water Sci. Technol., 18,* 63–89.

Dommergues, Y.R., Belser, L.W. and Schmidt, E.L. (1978) "Limiting factors for Microbial Growth and Activity in Soil," In *Advances in Microbial Ecology,* Ed., M. Alexander, *2,* 49.

Donaldson, W. (1932) "Some Notes on the Operation of Sewage Treatment Works," *Sewage Works J. 4,* 48.

Dondero, N.C. (1961) *"Sphaerotilus:* its Nature and Economic Significance," *Adv Appl. Microbiol., 3,* 77. Academic Press, New York, NY.

Dondero, N.C., Phillips, R.A. and Heukelekian, H. (1961) "Isolation and Preservation of Cultures of *Sphaerotilus,*" *Appl. Microbiol., 9,* 219.

Doohan, D. (1975) Chapter 7 "Rotifera" in *"Ecological Aspects of UsedWater Treatment."* Vol.I. "The Organisms and Their Ecology,"

Eds, Curds, C.R. and Hawkes, H.A., Academic Press, New York.

Eberhard, W.D. and Nesbitt, J.B. (1968) "Chemical Precipitation of Phosphorus in a High Rate Activated Sludge System," *J. Water Polln. Control Fedn., 40,* 1239.

Eikelboom, D.H. (1975) "Filamentous Organisms Observed in Bulking Activated Sludge," *Water Res., 9,* 365.

Eikelboom, D.H. (1977) "Identification of Filamentous Organisms in Bulking Activated Sludge," *Prog. Water Technol., 8,* 153.

Eikelboom, D.H. and van Buijsen H.J.J. (1981) *Microscopic Sludge Investigation Manual.* TNO Res. Inst. for Env. Hygiene, Delft, The Netherlands.

Eikelboom, D.H. (1982a) "Biosorption and Prevention of Bulking Sludge by Means of a High Floc Loading," Chapter 6 in *Bulking of Activated Sludge: Preventative and Remedial Methods,* Eds., B. Chambers and E.J. Tomlinson, Ellis Horwood Ltd, Chichester, England.

Eikelboom, D.H. (1982b) "Biological Characteristics of Oxidation Ditch Sludge," in *Oxidation Ditch Technology,* Proceedings published by CEP Consultants Ltd., Edinburgh, Scotland.

Ekama, G.A. and Marais, G.v.R. (1985) *Exploratory Study on Activated Sludge Bulking and Foaming Problems in South Africa (19831984),* Final Rept to Water Res. Commission, UCT Rept No. 54, Univ. of Cape Town, RSA.

Ekama, G.A. and Marais, G.v.R. (1986) "The Implication of the IAWPRC Hydrolysis Hypothesis on Low F/M Bulking," *Water Sci. Technol., 18,* 1119.

Farquhar, G.J. and Boyle, W.C. (1971a) "Identification of Filamentous Micro-organisms in Activated Sludge," *J. Water Polln. Control Fedn., 43,* 604.

Farquhar, G.J. and Boyle, W.C. (1971b) "Occurrence of Filamentous Micro-organisms in Activated Sludge," *J. Water Polln. Control Fedn., 43,* 779.

Farquhar, G.J. and Boyle, W.C. (1972) "Control of *Thiothrix* in Activated Sludge," *J. Water Polln. Control Fedn., 44,* 14.

Farrah, S.R., and Unz, R.F. (1976) "Isolation of Exocellular Polymer from *Zoogloea* Strains MP6 and 106 and from Activated Sludge," *Appl. Env. Microbiol., 32,* 33.

Faust, L. and Wolfe, R.S. (1961) "Enrichment and Cultivation of *Beggiatoa alba,*" *J. Bacteriol., 81,* 99.

Finger, R.E. (1973) "Solids Control in Activated

Sludge Plants With Alum," *J. Water Polln. Control Fedn., 45*, 1654.

Finstein, M.S. and Heukelekian, H. (1974) "Gross Dimensions of Activated Sludge Flocs with Reference to Bulking," *J. Water Polln Control Fedn., 39*, 33.

Frenzel, H.J. and Sarfert, F. (1971) "Erfahrungen über die Verhinderung von Blähschlammbildung durch Chlorung des belebten Schlammes," *Gas u. Wasserfach, 112*, 604.

Frenzel, H.J. (1977) "Some Experiences with the Chlorination of Activated Sludge for the Fight Against Bulking Sludge in the Berlin-Ruhleben Treatment Plant." *Prog. Water Technol., 8*, 163.

Gabb, D.M.D., Gonzalez, E. and Simons, W. (1985) "The Effects of a Selector on the Growth of Filamentous Bacteria in the Nitrification Activated Sludge Process," Internal Rept, Dept. of Water Polln. Control, City of San Jose, CA.

Gabb, D.M.D., Ekama, G.A., Jenkins, D. and Marais, G.v.R. (1987) "Specific Control Measures for Filamentous Bulking in Long Sludge Age Activated Sludge Systems," Res. Rept No. W61, Dept. of Civil Eng., Univ. of Capetown, RSA.

Gabb, D.M.D., Ekama, G.A., Jenkins, D. and Marais, G.v.R. (1989) "The Incidence of *Sphaerotilus natans* in Laboratory-Scale Activated Sludge Systems," *Water Sci. Technol., 21,* 29.

Gabb, D.M.D., Still, D.A., Ekama, G.A., Jenkins, D. and Marais, G.v.R. (1991) "The Selector Effect on Filamentous Bulking in Long Sludge Age Activated Sludge Systems," *Water Sci. Technol., 23*, 867.

Gaub Jr., W.H. (1924) "A Bacteriological Study of a Sewage Disposal Plant," *New Jersey Agr. Stat. Bull., No. 394 I*

Gaudy, E. and Wolfe, R.S. (1961) "Factors Affecting Filamentous Growth of *Sphaerotilus natans*," *Appl. Microbiol., 9*, 580.

Gaudy, E. and Wolfe, R.S. (1962) "Composition of an Extracellular Polysaccharide Produced by *Sphaerotilus natans*," *Appl. Microbiol., 10*, 200.

Gerardi, M.H. (1987) "An Operator's Guide to Free-Living Nematodes in Wastewater Treatment," *Public Works, 118*, 47.

Gerardi, M.H. (1987a) An Operator's Guide to Rotifers in Wastewater Treatment Processes," *Public Works, 118*, 66.

Goronszy M.C. (1979) "Intermittent Operation of the Extended Aeration Process for Small Systems," *J. Water Polln.Control Fedn., 51*, 274.

Goronszy, M.C. and Barnes, D. (1979) "Continuous Single Vessel Activated Sludge Treatment of Dairy Wastes," Proc. 87th Amer. Inst. Chem. Eng.Conf., Boston, MA.

Grau, P., Chudoba, J. and Dohanyos, M. (1982) "Theory and Practice of Accumulation-Regeneration Approach to the Control of Activated Sludge Filamentous Bulking," Chapter 7 in *Bulking of Activated Sludge: Preventative and Remedial Methods*, Eds. B. Chambers and E.J. Tomlinson, Ellis Horwood Ltd., Chichester, England.

Greenberg, A.E., Klein, G. and Kaufman,W.J. (1955) "Effect of Phosphorus on the Activated Sludge Process," *Sewage Ind. Wastes, 27*, 277.

Hale, F.D. and Garver, S.R. (1983) "Viscous Bulking of Activated Sludge," presented at the 56th Ann. Conf. Water Polln Control Fedn, Atlanta, GA, Oct. 2–7, 1983.

Hao, O.J. (1980), *"Preliminary Report on Nocardia Foaming,"* Internal Rept Sanitary Eng. and Env. Hlth Res. Lab., Univ. of Calif., Berkeley, CA.

Hao, O.J. (1982) "Isolation, Characterization and Continuous Culture Kinetics of a New *Sphaerotilus* Species Involved in Low Oxygen Activated Sludge Bulking," Ph.D. Dissertation, Dept. Civil Eng., Univ. of Calif., Berkeley, CA.

Hao, O.J., Richard, M.G., Jenkins, D., and Blanch, H. (1983) "The Half Saturation Coefficient for Dissolved Oxygen: A Dynamic Method for its Determination and its Effect on Dual Species Competition," *Biotechnol. Bioengrg., 35*, 403.

Hao, O.J., Strom, P.F. and Wu, Y.C. (1988) "A Review of the Role of *Nocardia*—like Filaments in Activated Sludge Foaming," *Water SA, 14*, 105.

Harold, R. and Stanier, R.Y. (1955) "The Genera *Leucothrix* and *Thiothrix*," *Bacteriol. Revs, 19*, 49.

Harris, F.W., Cockburn, T. and Anderson, T. (1927) "Biological and Physical Properties of Activated Sludge," *Water Works, 66*, 24.

Haseltine, T.R. (1932) "The Activated Sludge Process at Salinas, California, with Particular Reference to Cause and Control of Bulking," *Sewage Works J., 4*, 461.

Hattingh, W.H.J. (1963) "The Nitrogen and Phosphorus Requirement of Microorganisms," *Water and Waste Treatment, 10,* 380.

Hawkes, H.A. (1960) "Ecology of Activated Sludge and Bacteria Beds," in *Waste Treatment,* Ed., P.C.G. Isaac, Pergamon Press London.

Heide, B.A., and Pasveer, A. (1974), "Oxidation Ditch: Prevention and Control of Filamentous Sludge," *H₂0, 7,* 373.

Helmers, E.N., Frame, J.D., Greenberg, A.E., and Sawyer, C.N. (1952) "Nutritional Requirements in the Biological Stabilization of Industrial Wastes," *Sewage Ind. Wastes, 24,* 496.

Heukelekian, H. and Littman, M.L. (1939) "Carbon and Nitrogen Transformations in the Purification of Sewage by the Activated Sludge Process. I. With Mixtures of Sewage and Activated Sludge," *Sewage Works J., 11,* 226.

Heukelekian, H. and Littman, M.L. (1939) "II. Morphological and Biochemical Studies of Zoogloeal Organisms," *Sewage Works J., 11,* 752.

Heukelekian, H. and Ingols, R.S. (1940) "Studies on Activated Sludge Bulking. II. Bulking Induced by Domestic Sewage," *Sewage Works J., 12,* 694.

Heukelekian, H. (1941) "Activated Sludge Bulking," *Sewage Works J., 19,* 39.

Ho, C-F. and Jenkins, D. (1991) "The Effect of Surfactants on *Nocardia* Foaming in Activated Sludge," *Water Sci. Technol., 23,* 879.

Hoffman, J. (1987) "Influence of Oxic and Anoxic Mixing Zones in Compartment Systems on Substrate Removal and Sludge Characteristics in Activated Sludge Plants," *Water Sci. Technol., 19,* 897.

Hohnl, G. (1955) "Investigation into the Physiology and Nutrition of *Sphaerotilus natans,*" *Archiv. für Mikrobiol., 23,* 207.

Hong, S., Kisenbauer, K.S., Hartzog, D.G. and Fox, V.G. (1981) "A Biological Wastewater Treatment System for Nutrient Removal." Presented at 54th Ann. Conf. Water Polln Control Fedn, Detroit, MI.

Hong, S., Krichten, D., Best, A. and Kachwall, A. (1984) "Biological Phosphorus and Nitrogen Removal via the A/O Process: Recent Experience in the United States and United Kingdom," *Water Sci. Technol., 16,* 151172.

Houtmeyers, J. (1978) "Relations Between Substrate Feeding Pattern and Development of Filamentous Bacteria in Activated Sludge Processes," *Agricultura* (*Belgium*), *26,* 1.

Houtmeyers J., van den Eynde, E., Poffe, R., and Verachtert, H. (1980) "Relation Between Substrate Feeding Pattern and Development of Filamentous Bacteria in Activated Sludge Processes. I. Influence of Process Parameters," *European J. Appl. Microbiol. Biotechnol., 9,* 63.

Hünerburg, K., Sarfert, F. and Frenzel, H-J. (1970) "Ein Beitrag zum Problem Blähschlamm," *Gas. u. Wasserfach., 111,* 7.

Hutton, D.G. and Robertaccio, F.L. (1975) "Waste Water Treatment Process." US Patent 3,904,518. Sep. 9, 1975.

Hutton, D.G. and Robertaccio, F.L. (1978) "Industrial Waste Water Treatment Process." US Patent 4,069,148. Jan. 17, 1978.

Inamori, U., Takahashi, T. and Sudo, R. (1986) "Effects of Anaerobic Conditions on Activated Sludge Process," *J. Japan Sewage Works Assoc., 23,* 61–69.

Ingols, R.S. and Heukelekian, H. (1939) "Studies of Activated Sludge Bulking. I. Bulking of Activated Sludge by Means of Carbohydrates," *Sewage Works J., 11,* 927.

Ingols, R.S. and Heukelekian, H. (1940) "Studies on Activated Sludge Bulking. III. Bulking of Sludge Fed with Pure Substances and Supplied with Different Amounts of Oxygen," *Sewage Works J., 12,* 849.

Ip, S.Y., Bridger, J.S. and Mills, N.F. (1987) "Effects of Alternating Aerobic and Anaerobic Conditions on the Economics of the Activated Sludge System," *Water Sci. Technol., 19,* 911.

Jahn, T.L., Bovee, L. and Jahn, F.F. (1980) *"How to Know the Protozoa,"* 3rd Edn., Wm C.Brown Co., Dubuque, IA.

Javornicky, P., and Prokesova, V. (1963) "The Influence of Protozoa and Bacteria upon the Oxidation of Organic Substances in Water," *Int. Revue Ges., Hydrobiol. Hydogr. 48,* 335.

Jenkins, D., Neethling, J.B., Bode, H. and Richard, M.G. (1982) "Use of Chlorination for Control of Activated Sludge Bulking," Chapter 11, *Bulking of Activated Sludge: Preventative and Remedial Methods,"* Eds. B. Chambers and E.J. Tomlinson, Ellis Horwood Ltd., Chichester, England.

Jenkins, D., Parker, D.S., van Niekerk, A.M., Shao, Y-J and, Lee, S-E. (1983) "Relationship Between Bench Scale and Prototype Activated Sludge Systems," in *Scale-up of Water and Wastewater Treatment Processes,* Ed. N.W. Schmidtke, and D.W. Smith, Butterworth, 307.

Jenkins, D., Richard, M.G. and Neethling, J.B.

(1984) "Causes and Control of Activated Sludge Bulking," *Water Polln. Control, 83,* 455.

Jenkins, D. (1989), "Filaments Can Be Your Friends," Presented at 62nd Ann. Conf. *Water Polln Control Fedn.*, San Francisco, CA, Oct. 1989.

Johnstone, U.W.M., Rachwal, A.J. and Hanbury, M.J. (1979) "Settlement Characteristics and Settlement Tank Performance in Carrousel Activated Sludge Systems," *Water Polln. Control, 78,* 337.

Jones, P.H. (1964) "Studies of the Ecology of the Filamentous Sewage Fungus, *Geotrichum candidum*," Ph.D. Dissertation Northwestern Univ., Evanston, IL.

Keefer, C.E. (1963) "Relationship of Sludge Density Index to the Activated Sludge Process," *J. Water Polln. Control Fedn., 35,* 1166.

Keinath, T.M. (1977) "Activated Sludge Unified Design and Operation," *J. Env. Eng. Div. Amer. Soc. Civil Eng., 103,* 829.

Keller, P.J. and Cole, C.A. (1973) "Hydrogen Peroxide Cures Bulking," *Water and Wastes Eng., 10,* E4.

Koopman, B.L., Lau, A.O., Strom, P.F. and Jenkins, D. (1980) "A Simple Device for Level Control With Subsurface Drawoff in Chemostats," *Biotechnol. Bioeng., 22,* 2433.

Koopman, B. and Cadee, K. (1983) "Prediction of Thickening Capacity Using Diluted Sludge Volume Index," *Water Res., 17,* 1427.

Koopman, B.L. Lau, A.O., Strom, P.F. and Jenkins, D. (1983) "Liquid Level Control by Subsurface Drawoff," U.S. Patent 4,370,418, Jan. 15, 1983.

Kraus, L.S. (1945) "The Use of Digested Sludge and Digestor Overflow to Control Activated Sludge Bulking," *Sewage Works J., 170,* 1177.

Krinsky, N.I. (1979) "Carotenoid Pigments: Multiple Mechanisms For Coping With Stress of Photosensitized Oxidations," in *Strategies of Microbial Life in Extreme Environments.* Ed. M. Shilo, 163 Verlag. Weinheim.

Kroiss, H. (1985) "Bulking Problems in the Leopoldsdorf Sugar Mill Plant," *Wiener Mitteilungen Band 56,* 81.

Kroiss, H. and Ruider, E. (1977) "Comparison of the Plug-Flow and Complete-Mix Activated Sludge Process," *Prog. Water Technol., 8,* 169.

Krul, J.M. (1977) "Experiments with *Haliscomenobacter hydrossis* in Continuous Culture Without and With *Zoogloea ramigera*," *Water Res., 11,* 197.

Kucman, K. (1987) "Activated Sludge Process Combined with Biofilm Cultivation," Ph.D. Dissertation, Inst. Chem. Technol., Prague, Czechoslovakia.

Lackey, J.B. and Wattie, E. (1946) "The Biology of *Sphaerotilus natans* Kutzing in Relation to the Bulking of Activated Sludge," *Sewage Works J., 12,* 669.

Lakay, M.T., Wentzel, M.C., Ekama, G.A. and Marais, G.v.R. (1988) "Bulking Control With Chlorination in a Nutrient Removal Activated Sludge System," *Water SA, 14,* 35.

Larkin, J.M. (1980) "Isolation of *Thiothrix* in Pure Culture and Observation of a Filamentous Epiphyte on *Thiothrix*," *Current Microbiol., 4,* 155.

Larkin, J.M. and Strohl, W.R. (1983) "*Beggiatoa, Thiothrix and Thioplica*," *Ann. Rev. Microbiol., 37,* 341.

Lau, A.O., Strom, P.F. and Jenkins, D. (1984a) "Growth Kinetics of *Sphaerotilus natans* and a Floc Former in Pure and Dual Continuous Culture," *J. Water Polln. Control Fedn., 56,* 41.

Lau, A.O., Strom P.F. and Jenkins, D. (1984b) "The Competitive Growth of Floc-forming and Filamentous Bacteria: a Model for Activated Sludge Bulking," *J. Water Polln. Control Fedn., 56,* 52.

Lechevalier, M.P. and Lechevalier, H.A. (1974) "*Nocardia amarae* sp. nov., an Actinomycete Common in Foaming Activated Sludge," *Inter.J. Syst. Bacteriol., 24,* 278.

Lechevalier, H.A. (1975) *Actinomycetes of Sewage Treatment Plants*, USEPA Rept No. 600/2-75/031, Cincinnati, OH.

Lechevalier, H.A., Lechevalier, M.P. and Wyszkowski, P.E. (1977) *Actinomycetes of Sewage Treatment Plants*, EPA Report 600/2-77-145, Cincinnati, OH.

Lee, E.G.H., Mueller, J.C. and Walden, C.C. (1975) "Effect of Temperature and Sludge Loading on BOD_5 Removal and Sludge Settleability in Activated Sludge Systems Treating Bleached Kraft Effluents," *TAPPI J. 58,* 100.

Lee, S-E., Koopman, B.L., Jenkins, D., and Lewis, R.F. (1982) "The Effect of Aeration Basin Configuration on Activated Sludge Bulking at Low Organic Loading," *Water Sci. Technol., 14,* 407.

Lee, S-E., Koopman, B.L., Bode, H., and Jenkins, D. (1983) "Evaluation of Alternative Sludge Settleability Indices," *Water Res., 17,* 1421.

Lemmer, H. (1986) "The Ecology of Scum Causing Actinomycetes in Sewage Treatment Plants," *Water Res., 20,* 531.

Lemmer, H. and Kroppenstedt, R.M. (1984) "Chemotaxonomy and Physiology of Some Actinomycetes Isolated from Scumming Activated Sludge," *System. Appl. Microbiol., 5,* 124.

MacDonald, C.R., Cooper, D.G. and Zajic, J.E. (1981) "Surface Active Lipids from *Nocardia erythropolis* Grown on Hydrocarbons," *Appl. Env. Microbiol., 41,* 117.

Mamais, D. and Jenkins, D. (1992) "The Effects of MCRT and Temperature on Enhanced Biological Phosphorus Removal." Presented at 16th Intl Assoc. for Water Polln. Res. and Control, Washington DC, USA., May 1992.

Margaritis, A., Kennedy, K., Zajic, J.E., Gerson, D.F. (1979) "Biosurfactant Produced by *Nocardia erythropolis*," *Dev. Ind. Microbiol., 20,* 623.

Marshall, K.C. and Cruickshank, R.H. (1973) "Cell Surface Hydrophobicity and Orientation of Certain Bacteria at Interfaces," *Arch. Mikrobiol., 91,* 29.

Matsche, N.F. (1977) "Blähschlamm-Versuche und Bekämpfung. Wiener Mittelunger," *Wasser Abwasser Gewasser, 22,* 1.

Matsche, N. (1980) "Gutachten uber die Bakteriologische Zusammensetzung des Belebtschlammes in der Kläranlage Wien-Blumenthal," Inst. für Wassersorgung, Abwassereinigung und Gewässerschutz der Technischen Universität Wien., Vienna, Austria.

Matsche, N.F. (1982) "Control of Bulking Sludge—Practical Experiences in Austria," *Water Sci. Technol. 14,* 311.

Matsui, S. and Yamamoto, R. (1983) "Measuring the Lengths of Filamentous Microbes in Activated Sludges using a Video Color TV with a Microscope," *Water Sci. Technol., 16:10–11,* 69.

McKinney, R.E. and Horwood, M.P. (1952) "Fundamental Approach to the Activated Sludge Process. I. Floc-Producing Bacteria," *Sewage Ind. Wastes, 24,* 117.

McKinney, R.E. and Weichlein, R.G. (1953) "Isolation of Floc-Producing Bacteria from Activated Sludge," *Appl. Microbiol., 1,* 259.

McKinney, R.E. (1955) *Bacterial Flocculation in Relationship to Aerobic Waste Treatment Processes.* Nat. Inst. Hlth, Prog. Rept, Project. E-645, MIT, Cambridge, MA.

McKinney, R.E. (1957) "Activity of Microorganisms in Organic Waste Disposal. II. Aerobic Processes," *Appl. Microbiol., 5,* 167.

Merkel, G.J. (1975) "Observations on the Attachment of *Thiothrix* to Biological Surfaces in Activated Sludge," *Water Res., 9,* 881.

Minnikin, D.E. (1982) in *"The Biology of the Mycobacteria" Volume 1.* Eds, C. Ratledge and J. Stanford, Academic Press, London, 95–184.

Miyamoto-Mills, J., Larson, J., Jenkins, D. and Owen, W.F. (1983) "Design and Operation of a Pilot-Scale Biological Phosphate Removal Plant at Central Contra Costa Sanitary District, CA." *Water Sci. Technol. (Capetown), 15,* 153.

Mohlman, F.W. (1934) "The Sludge Index," *Sewage Works J. 6,* 119.

Mudrack, K. and Kunst, S. (1986) *"Biology of Sewage Treatment and Water Pollution Control,"* John Wiley and Sons, NY.

Mulder, E.G. (1964) "Iron Bacteria, Particularly Those of the *Sphaerotilus-Leptothrix* Group, and Industrial Problems," *J. Appl. Bacteriol., 27,* 151.

Mulder, E.G. and van Veen, W.L. (1963) "Investigations on the *Sphaerotilus-Leptothrix* Group," *Antonie van Leeuwenhoek, 29,* 121.

Neethling, J.B., Johnson, K.M., and Jenkins D. (1982) Chemical and Microbiological Aspects of Filamentous Bulking Control Using Chlorination, UCB-SEEHRL Rept No. 82–2, Sanitary Eng. Env. Hlth Res. Lab., Univ. of Calif., Berkeley, CA.

Neethling, J.B. (1984) "The Control of Activated Sludge Bulking by Chlorination," Ph.D. Dissertation, Dept. Civil Eng., Univ. of Calif., Berkeley, CA.

Neethling, J.B., Jenkins, D. and Johnson, K.M. (1985a) "Chemistry, Microbiology and Modelling of Chlorination for Activated Sludge Bulking Control," *J. Water Polln. Control Fedn, 57,* 882.

Neethling, J.B., Johnson, K.M. and Jenkins, D. (1985b) "Using ATP to Determine the Chlorine Resistance of Filamentous Bacteria Associated with Activated Sludge Bulking," *J. Water Polln. Control Fedn., 57,* 890.

Neethling, J.B., Chung, Y-C. and Jenkins, D. (1987) "Activated Sludge Chlorine Reactions During Bulking Control," *J. Env. Eng. Div. Amer. Soc. Civil Eng., 113,* 136.

Nielsen, P.H. (1985), "Oxidation of Sulfide and Thiosulfate and Storage of Sulfur Granules in *Thiothrix* from Activated Sludge," *Water Sci. Technol., 17,* 167.

Norris, D.P., Parker, D.S., Daniels, M.S. and Owens, E.L. (1982) "Production of High

Quality Trickling Filter Effluent Without Tertiary Treatment," *J. Water Polln. Control Fedn., 54*, 1087.

Palm, J.C., Jenkins, D., and Parker, D.S. (1980) "Relationship Between Organic Loading, Dissolved Oxygen Concentration and Sludge Settleability in the Completely-Mixed Activated Sludge Process," *J. Water Polln. Control Fedn., 52*, 2484.

Parker, D.S., Jenkins, D. and Kaufman, W.J. (1971) "Physical Conditioning of the Activated Sludge Floc," *J. Water Polln. Control Fedn., 43*, 1817.

Parker, D.S., Jenkins, D. and Kaufman, W.J. (1972) "Floc Breakup in Turbulent Flocculation Processes," *J. Sanitary Eng. Div., Amer. Soc.Civil Eng., 98*, SAI, 79.

Patoczka, J. and Eckenfelder, W.W. (1990) "Performance and Design of a Selector for Bulking Control," *Res. J. Water Polln. Control Fedn, 62*, 151.

Pavoni, J.L., Tenney, M.W. and Echelberger, W.T. Jr. (1972) "Bacterial Exocellular Polymers and Biological Flocculation," *J. Water Polln. Control Fedn., 44*, 414.

Pearse, L. and APHA Committee (1937) "Bulking of Sludge in the Activated Sludge Process of Sewage Treatment," *APHA Year Book, 27*, 164.

Peptiprez, M. and Leclerc, H. (1969) "Les Bacteries du Group *Sphaerotilus-Leptothrix*," *Ann. Inst. Pasteur Lille, 20*, 115.

Phaup, J.D. (1968) "The Biology of *Sphaerotilus* Species," *Water Res., 2*, 597.

Pike, E.B. (1972) "Aerobic Bacteria," in *Ecological Aspects of Used Water Treatment*, 1. Eds, C.R. Curds and H.A. Hawkes, Academic Press, N.Y. 1–63.

Pipes, W.O. and Jones, P.H. (1963) "Decomposition of Organic Wastes by *Sphaerotilus*," *Biotech. Bioeng., 5*, 287.

Pipes, W.O. (1974) "Control Bulking with Chemicals," *Water Wastes Eng., 9*, 30.

Pipes, W.O. (1978a) "Microbiology of Activated Sludge Bulking," *Adv. Appl. Microbiol., 24*, 85.

Pipes, W.O. (1978b) "Actinomycete Scum Production in Activated Sludge Processes," *J. Water Polln. Control Fedn., 50*, 628.

Pipes, W.O. (1979) "Bulking, Deflocculation and Pin-Point Floc," *J. Water Polln. Control Fedn., 51*, 62.

Pitman, A.R. (1985) "Settling of Nutrient Removal Activated Sludges," *Water Sci. Technol., 17*, 493.

Pitt, P.A. and Jenkins, D. (1990) "Causes and Control of *Nocardia* in Activated Sludge," *Res. J. Water Polln. Control Fedn., 62*, 143.

Poffe, R., van der Leyden J., and Verachtert, H. (1979) "Characterization of a *Leucothrix*-Like Bacterium Causing Sludge Bulking During Petrochemical Waste Water Treatment," *European J. Microbiol. Biotechnol., 8*, 229.

Pretorius, W.A. (1987) "A Conceptual Basis for Microbial Selection in Biological Waste Treatment," *Water Res., 21*, 891.

Pretorius, W.A. and Laubscher, C.J.P. (1987) "Control of Biological Scum in Activated Sludge Plants by Means of Selective Flotation," *Water Sci. Technol., 19*, 1003.

Price, G.J. (1982) "Use of an Anoxic Zone to Improve Activated Sludge Settleability," Supplementary Contribution No (iii) in *Bulking of Activated Sludge: Preventive and Remedial Methods*, Eds, B. Chambers and E.J. Tomlinson, Ellis Horwood Ltd., Chichester, England, 259.

Pringsheim, E.G. (1949) "The Filamentous Bacteria *Sphaerotilus, Leptothrix, Cladothrix* and their Relation to Iron and Manganese," *Phil. Trans. Roy. Soc. London, 233B*, 453.

Pringsheim, E.G.(1951) "The *Vitreoscilla*: A Family of Colorless, Gliding Filamentous Organisms," *J. Gen. Microbiol., 5*, 124.

Pringsheim, E.G. and Wiessner, W. (1963) "Minimum Requirements for Heterotrophic Growth and Reserve Substances in *Beggiatoa*," *Nature (London), 197*, 102.

Pringsheim, E.G. (1964) "Heterotrophism and Species Concept in *Beggiatoa*," *Amer. J. Botany, 51*, 898.

Rachwal, A.J., Johnstone, D.W.M., Hanbury, M.J., and Critchard, D.J. (1982) "The Application of Settleability Tests for the Control of Activated Sludge Plants," Chapter 13 in *Bulking of Activated Sludge: Preventative and Remedial Methods*, Eds, B. Chambers and E.J. Tomlinson, Ellis Horwood Ltd., Chichester, England.

Reid, J.R. (1991) "Phosphorus and Filaments" *TAPPI Environ. Conf. Proc., 1*, 99.

Rensink, J.H. (1974) "New Approach to Preventing Bulking Sludge," *J. Water Polln. Control Fedn., 46*, 1888.

Rensink, J.H., Jellema, K. and Ywema, T. (1977) "The Influence of Substrate Gradient on the Development of Bulking Sludge," *H₂O, 10*, 338.

Rensink, J.H., Leentvaar, J. and Donker, H.J.G.W. (1979) "Control of Bulking Combined with Phosphate Removal by Ferrous Sulphate Dosing," *H₂O, 12,* 150.

Rensink, J.H., Donker, H.J.G.W. and Ijwema, T.S.J. (1982) "The Influence of Feed Pattern on Sludge Bulking," Chapter 9 in *Bulking of Activated Sludge: Preventative and Remedial Methods*, Eds, B. Chambers and E.J. Tomlinson, Ellis Horwood Ltd., Chichester, England.

Rensink, J.H. and Donker, H.J.G.W. (1987) "The Influence of Bulking Sludge on Enhanced Biological Phosphorus Removal," In *Biological Phosphate Removal From Wastewaters*, Ed., R. Ramadori, Pergamon Press, 369.

Reynoldson, T.B., (1942) "*Vorticella* as an Indicator Organism in Activated Sludge," *Nature, (London), 149,* 608.

Richard, M.G., Jenkins, D. Hao, O., and Shimizu, G. (1982a) *The Isolation and Characterization of Filamentous Micro-organisms from Activated Sludge Bulking*," Rept No. 81-2, Sanitary Eng. Env. Hlth Res. Lab., Univ. of Calif., Berkeley, CA.

Richard, M.G., Shimizu, G., Jenkins, D., Williams, T. and Unz. R.F. (1983) "Isolation and Characterization of *Thiothrix* and *Thiothrix*-Like Filamentous Organisms from Bulking Activated Sludge," presented at the 1983 Ann. Mtg, Amer. Soc. for Microbiol., New Orleans, LA (Abs. Ann. Meet., Q60, 270).

Richard, M.G., Hao, O. and Jenkins, D. (1985) "Growth Kinetics of *Sphaerotilus* Species and their Significance in Activated Sludge Bulking," *J. Water Polln. Control Fedn., 57,* 64.

Richard, M.G. and Jenkins, D. (1985), "The Causes and Control of Activated Sludge Bulking," *TAPPI Journal, 68,* 73.

Richard, M.G., Shimizu, G.P., and Jenkins, D. (1985) "The Growth Physiology of the Filamentous Organism Type 021N and its Significance to Activated Sludge Bulking," J. *Water Polln. Control Fedn., 57,* 1152.

Richard, M.G. (1986) "Filamentous Microorganisms and Activated Sludge Operation in Colorado: Operational Problems, Diagnosis and Remedial Measures." Presented at Rocky Mountain Water Polln Assoc. Ann. Mtg, Breckenridge, CO.

Richard, M.G. (1989) "*Activated Sludge Microbiology*," Water Polln Control Fedn, Alexandria, VA, 73pp.

Richard, M.G. (1991) "Filamentous and Polysaccharide Bulking Problems and Their Control in Papermill Activated Sludge," *TAPPI Env. Conf. Proc., 1,* 95.

Richards, T., Nungesser, P. and Jones, C. (1990) "Solution of *Nocardia* Foaming Problems," *Res. J. Water Polln. Control Fedn., 62,* 915.

Ridenaur, G.M., Henderson, C.N. and Schulhoff, H.B. (1937) "Operation Experiences With Chlorination of Activated Sludge," *Sewage Works J., 9,* 63.

Rouf, M.A. and Stokes. J.L. (1964) "Morphology, Nutrition and Physiology of *Sphaerotilus discophorus*," *Archiv für Mikrobiol., 49,* 132.

Ruchhoft, C.C. and Kachmar, J.F. (1941) "Studies of Sewage Purification. XIV. The Role of *Sphaerotilus natans* in Activated Sludge Bulking," *Sewage Works J., 13,* 3.

Salameh, M.F. and Malina, J.F. Jr. (1989) "The Effects of Sludge Age and Selector Configuration on the Control of Filamentous Bulking in the Activated Process," *J. Water Polln. Control Fedn., 61,* 1510.

Schiemer, F. (1975) Chapter 6 "Nematoda" in "*Ecological Aspects of Used-Water Treatment. Vol.I. The Organisms and Their Ecology*," Eds, C.R. Curds, and H. A. Hawkes, Academic Press, New York.

Scotten, H.L. and Stokes, J.L. (1962) "Isolation and Properties of *Beggiatoa*," *Archiv f-r Mikrobiol., 42,* 353.

Sedlak, R.I. editor (1991) "*Phosphorus and Nitrogen Removal from Municipal Wastewater: Principles and Practice*." 2nd edn. Lewis Publishers, Chelsea, MI.

Segerer, M. (1984) "Untersuchungen zur Schwimmschlammbildung in Kläranlagen durch Actinomyceten," *Korr. Abwasser, 12* 1073.

Seviour, E.M., Williams, C.J., Seviour, R.J., Soddrell, J.A. and Lindrea, K.C. (1990) "A Survey of Filamentous Bacterial Populations From Foaming Activated Sludge Plants in Eastern States of Australia," *Water Res., 24,* 493.

Sezgin, M. (1977) "The Effect of Dissolved Oxygen Concentration on Activated Sludge Process Performance," Ph.D. Dissertation, Univ. of Calif., Berkeley, CA.

Sezgin, M., Jenkins, D, and Parker, D.S. (1978) "A Unified Theory of Filamentous Activated Sludge Bulking," *J. Water Polln. Control Fedn., 50,* 362.

Sezgin, M., Palm, J.H. and Jenkins, D.(1980) "The Role of Filamentous Micro-organisms in Activated Sludge Settling," *Prog. Water Technol., 12,* 171.

Sezgin, M. and Karr, P.R. (1986) "Control of

Actinomycete Scum on Aeration Basins and Clarifiers," *J. Water Polln. Control Fedn., 58,* 972.

Shao, Y-J. and Jenkins, D. (1989) "The Use of Anoxic Selectors for the Control of Low F/M Activated Sludge Bulking," *Water Sci. Technol., 21,* 609.

Singer, P.C., Pipes, W.O. and Herman, E.R. (1968) "Flocculation of Bulked Activated Sludge With Polyelectrolytes," *J. Water Polln. Control Fedn., 40,* Part 2, R1.

Sladka, A. and Ottova, V. (1973) "Filamentous Organisms in Activated Sludge," *Hydrobiologia, 43,* 285.

Slijkhuis, H. and Deinema, M.H. (1982) "The Physiology of *Microthrix parvicella*, a Filamentous Bacterium Isolated from Activated Sludge," Chapter 5 in *Bulking of Activated Sludge: Preventative and Remedial Methods,* Eds, B. Chambers and E.J. Tomlinson, Ellis Horwood Ltd., Chichester, England.

Slijkhuis, H. (1983) "The Physiology of the Filamentous Bacterium *Microthrix parvicella*," Ph.D. Dissertation, Wageningen, Holland.

Slijkhuis, H. (1983a) "*Microthrix parvicella*, a Filamentous Bacterium Isolated from Activated Sludge: Cultivation in a Chemically Defined Medium," *Appl. Env. Microbiol., 46,* 832.

Slijkhuis, H., van Groenestijn, J.W. and Kijlstra, D.J. (1984) "*Microthrix parvicella*, a Filamentous Bacterium from Activated Sludge: Growth on Tween 80 as Carbon and Energy Source," *J. Gen. Microbiol., 130,* 2035.

Slijkhuis, H. and Deinema, M.H. (1988) "Effect of Environmental Conditions on the Occurrence of *Microthrix parvicella* in Activated Sludge," *Water Res., 22,* 825.

Smit, J. (1932) "A Study of the Conditions Favoring Bulking of Activated Sludge," *Sewage Works J., 4,* 960.

Smit, J. (1934) "Bulking of Activated Sludge. II. Causative Organisms." *Sewage Works J., 6,* 1039.

Smith, R.S. and Purdy, W.C. (1936) "Studies of Sewage Purification, IV. The Use of Chlorine for the Correction of Sludge Bulking in the Activated Sludge Process," *Public Hlth Repts, 51,* 617.

Soddell, J.A. and Seviour, R.J. (1990) "Microbiology of Foaming in Activated Sludge Plants." *J. Appl. Bacteriol., 69,* 145.

Spector, M.L. (1977) "Production of Non-Bulking Activated Sludge," U.S. Patent 4,056,465. Reissue 32,429. Jun. 2, 1987.

Spock, B. (1979) "The Social Benefits of a Filament-Free Childhood," *Bezerkley Press, 36,* 24–36.

Standard Methods for the Examination of Water and Wastewater, 17th edn (1991). *Amer. Publ. Hlth. Assoc.,* New York, NY.

Steytler, R.B., Goddard, M.F., Ambrose, W.A. and Wilson, T.E. (1981) "Listening to the Activated Sludge Plant in Phoenix," presented at 54th Ann. Conf., Water Polln. Control Fedn., Detroit, MI.

Steytler, R.B., Goddard, M.F., Buhr, H.O., and Wilson T.E. (1982) "Watching the 91st Avenue Plant in Phoenix," presented at 55th Ann. Mtg, Arizona Water Polln. Control Assoc., Flagstaff, AZ.

Still, D., Blackbeard, J.R., Ekama, G.A. and Marais, G.v.R. (1986) "The Effect of Feeding Patterns on Sludge Growth Rate and Sludge Settleability," Res. Rept No. W55, Dept. Civil Eng., Univ. of Capetown, RSA.

Stobbe, G. (1964) "Über das Verhalten von Belebtschramm in aufsteigender Wasserbewegung," Veröffentlichungen des Institutes für Siedlungwasserwirtschaft der Technischen Hochschule Hannover, *18.*

Stokes, J.L. (1954) "Studies on the Filamentous Sheathed Bacterium *Sphaerotilus natans*," *J. Bacteriol., 67,* 278.

Strohl, W.R. and Larkin, J.M. (1978) "Enumeration, Isolation and Characterization of *Beggiatoa* from Freshwater Sediments," *Appl. Env. Microbiol., 36,* 755.

Strom, P.F. and Finstein, M.S. (1977) "Nitrification in a Chlorinated Activated Sludge Culture," *J. Water Polln. Control Fedn., 49,* 584.

Strom, P.F. and Jenkins, D. (1984) "Identification and Significance of Filamentous Microorganisms in Activated Sludge," *J. Water Polln. Contol Fedn., 56,* 52.

Strunk, W.G. and Shapiro, J. (1976) "Bulking Control Made Easy with Hydrogen Peroxide." *Water Polln. Control, 114,* 10.

Tabor, W.A. (1976) "Wastewater Microbiology," *Ann. Rev. Microbiol., 30,* 263.

Tago, Y. and Aiba, K. (1977) "Exocellular Mucopolysaccharide Closely Related to Bacterial Floc Formation," *Appl. Env. Microbiol., 34,* 308.

Tapleshay, J.A. (1945) "Control of Sludge Index by Chlorination of Return Sludge," *Sewage Works J., 17,* 1210.

Tarjan, A.C., Esser, R.P. and Chang, S.L. (1977) "An Illustrated Key to Nematodes Found in

Fresh Water," *J. Water Polln. Control Fedn.* *49*, 2318.

Thirion, N. (1983) "Chlorination as a Control Measure for Bulking Sludge at the Daspoort Sewage Treatment Plant, Pretoria" *SA Water Bulletin*, Water Res. Commission, Pretoria, RSA, 13.

Thomanetz, E. and Bardtke, D. (1977) "Studies on the Possibility for Control of Bulking Sludge with Synthetic Flocculation Agents," *Korr. Abwasser, 1824*, 15.

Tomlinson, E.J. (1976) "Bulking—A Survey of Activated Sludge Plants," Tech. Rept TR35, Water Res. Centre, Stevenage, England.

Tomlinson, E.J. (1978) "The Effect of Anoxic Mixing Zones upon the Settleability of Activated Sludge" Unpublished Rept, Water Res. Centre, Stevenage, England.

Tomlinson, E.J. and Chambers, B. (1978a) "The Use of Anoxic Mixing Zones to Control the Settleability of Activated Sludge," Tech. Rept TR116, Water Res. Centre, Stevenage, England.

Tomlinson, E.J. and Chambers, B. (1978b) "The Effect of Longitudinal Mixing on the Settleability of Activated Sludge," Tech. Rept TR 122, Water Res. Centre, Stevenage, England.

Tomlinson, E.J. and Chambers, B. (1979) "Methods for Prevention of Bulking in Activated Sludge," *Water Polln. Control, 78*, 524.

Tomlinson, E.J. (1982) "The Emergence of the Bulking Problem and the Current Situation in the UK," Chapter 1 in *Bulking of Activated Sludge: Preventative and Remedial Methods*, Eds, B. Chambers and E.J. Tomlinson, Ellis Horwood Ltd., Chichester, England.

Torpey, W.N. (1948) "Practical Results of Step Aeration," *Sewage Works J., 20*, 781.

Torpey, W.N. and Chasick, A.H. (1955) "Principles of Activated Sludge Operation," *Sewage Works J., 27*, 1217.

Tucek, F. and Chudoba, J. (1969) "Purification Efficiency in Aeration Tanks with Complete Mixing and Piston Flow," *Water Res., 3*, 559.

Unz, R. (1979) Privately communicated, Dept. of Civil Eng, The Pennsylvania State Univ., State College, PA.

Unz, R.F. and Farrah, S.R. (1976) "Exopolymer Production and Flocculation by *Zoogloea* MP6," *Appl. Env. Microbiol., 31*, 623.

van den Eynde, E., Vriens, L., and Verachtert, H. (1982a) "Relation between Substrate Feeding Pattern and Development of Filamentous Bacteria in Activated Sludge Processes. III. Applications with Industrial Waste Waters,"

European J. Appl. Microbiol. Biotechnol., 15, 246.

van den Eynde, E., Houtmeyers, J., and Verachtert, H. (1982b) "Relationship between Substrate Feeding Pattern and Development of Filamentous Bacteria in Activated Sludge," Chapter 8 in *Bulking of Activated Sludge: Preventative and Remedial Methods*, Eds. B. Chambers and E.J. Tomlinson, Ellis Horwood Ltd., Chichester, England.

van den Eynde, E., Geerts, J., Maes B., and Verachtert, H. (1983) "Influence of the Feeding Pattern on the Glucose Metabolism of *Arthrobacter* sp. and *Sphaerotilus natans*, Growing in Chemostat Culture, Simulating Activated Sludge Bulking," *European J. Appl. Microbiol. Biotechnol., 17*, 35.

van den Eynde, E. (1983) "Substrate Stress: A Condition for Safe Operation of Activated Sludge Systems with Regard to Filamentous Bulking," Ph.D. Dissertation, Katholieke Universiteit de Leuven, Belgium.

van den Eynde, E., Wynants, M., Vriens, L., Verachtert, H. (1984) "Transient Behavior and Time Aspects of Intermittently and Continuously Fed Bacterial Cultures with Regard to Filamentous Bulking of Activated Sludge," *Appl. Microbiol. Biotechnol., 19*, 44.

van den Eynde, E., Vriens, L., De Cuyper, P. and Verachtert, H. (1984) "Plug Flow Simulating and Completely Mixed Reactors with a Premixing Tank in the Control of Filamentous Bulking," *Appl. Microbiol. Biotechnol., 19*, 288.

van Leeuwen, J. and van Rossum, P.G. (1990) "Trihalomethane Formation during Bulking Control with Chlorine." *J. Inst. Water Env. Management, 4*, 530.

van Niekerk, A.M. (1985) "Competitive Growth of Flocculant and Filamentous Microorganisms in Activated Sludge Systems," Ph.D. Dissertation, Dept. Civil Eng., Univ. of Calif. Berkeley, CA.

van Niekerk, A.M., Malea, A., Kawahigashi, J., and Reichlin, D. (1987) "Foaming in Anaerobic Digesters: A Survey and Laboratory Investigation," *J. Water Polln. Control Fedn., 59*, 249.

van Niekerk, A.M., Richard, M.G. and Jenkins, D. (1987) "The Competitive Growth of *Zoogloea ramigera* and Type 021N in Activated Sludge and Pure Culture—A Model for Low F/M Bulking," *J. Water Polln. Control Fedn., 59*, 262.

van Niekerk, A.M., Jenkins, D. and Richard,

M.G. (1988) "A Mathematical Model of the Carbon-Limited Growth of Filamentous and Floc Forming Organisms in Low F/M Sludge," *J. Water Polln. Control Fedn., 60,* 100.

van Veen, W.L. (1972) "Factors Affecting the Oxidation of Manganese by *Sphaerotilus discophorus*," *Antonie van Leeuwenhoek, 38,* 623.

van Veen, W.L. (1973) "Bacteriology of Activated Sludge, in Particular the Filamentous Bacteria," *Antonie van Leeuwenhoek, 39,* 189.

van Veen, W.L., van der Eooij, K., Gruze, E.C.W.A., and van der Vlies, A.W. (1973) "Investigations of the Sheathed Bacterium *Haliscomenobacter hydrossis*, a Bacterium Occurring in Bulking Activated Sludge," *Antonie van Leeuwenhoek, 39,* 207.

van Veen, W.L., Krul, J.M., and Bulder, C.J.E.A. (1982) "Some Growth Parameters of *Haliscomenobacter hydrossis*, a Bacterium Occurring in Bulking Activated Sludge," *Water Res., 16,* 531.

Vega-Rodriquez, B.A. (1983) "Quantitative Evaluation of *Nocardia* spp. Presence in Activated Sludge," M.S. Thesis, Dept. Civil Eng., Univ. of Calif., Berkeley, CA.

Verachtert, H., van den Eynde, E., Poffe, R. and Houtmeyers, J. (1980) "Relation between Substrate Feeding Pattern and Development of Filamentous Bacteria in Activated Sludge Processes. II. Influence of Substrates Present in the Influent," *European J. Appl. Microbiol. Biotechnol., 9,* 137.

Wagner, F. (1982) "Study of the Causes and Prevention of Sludge Bulking in Germany," Chapter 2 in *Bulking of Activated Sludge: Preventative and Remedial Methods*, Eds, B. Chambers and E.J. Tomlinson, Ellis Horwood Ltd., Chichester, England.

Waitz, S. and Lackey J.B. (1959) "Morphological and Biochemical Studies on the Organism *Sphaerotilus natans*," *Qtrly. J. Fla. Acad. Sci., 21,* 335.

Wanner, J., Ottova, V. and Grau, P. (1987) "Effect of an Anaerobic Zone on Settleability of Activated Sludge," In: *Biological Phosphate Removal From Wastewaters*, Ed., R. Ramadori, Pergamon Press, 155.

Wanner, J., Chudoba, J., Kucman, K. and Proske, L. (1987a) "Control of Activated Sludge Filamentous Bulking—VII Effect of Anoxic Conditions," *Water Res., 21,* 1447.

Wanner, J., Kucman, K., Ottova, V. and Grau, P. (1987b) "Effect of Anaerobic Conditions on Activated Sludge Filamentous Bulking in Laboratory System," *Water Res., 21,* 1541.

Wanner, J. and Grau, P. (1988) "Filamentous Bulking in Nutrient Removal Activated Sludge Systems," *Water Sci. Technol., 20,* 1.

Wanner, J. and Grau, P. (1989) "Identification of Filamentous Microorganisms from Activated Sludge: a Compromise Between Wishes, Needs and Possibilities," *Water Res., 23,* 883.

Wanner, J. and Novak, L. (1990) "The Influence of a Particulate Substrate on Filamentous Bulking and Phosphorus Removal in Activated Sludge Systems," *Water Res., 24,* 573.

Watanabe, A., Miya, A. and Matsuo, Y. (1984) "Laboratory Scale Study on Biological Phosphate Removal Using Synthetic Wastewater," Newsletter of the IAWPRC Specialist Group on Phosphate Removal in Biological Sewage Treatment Processes, 2, No. 1, 40.

Weddle, C.L. and Jenkins, D. (1971) "The Viability and Activity of Activated Sludge," *Water Res., 5,* 621.

Wiechers, H.N.S. (1983) "Sludge Bulking in Heidelberg," *SA Water Bulletin*, Water Res. Commission, Pretoria, RSA, 11.

Wells, W.N. and Garret, M.T. Jr. (1971) "Getting the Most from an Activated Sludge Plant," *Public Works, 102,* 63.

Wheeler, D.R. and Rule, A.M. (1980) "The Role of *Nocardia* in the Foaming of Activated Sludge: Laboratory Studies," presented at Georgia Water and Polln. Control Assoc. Ann. Mtg, Savannah, GA.

Wheeler, M., Jenkins, D., and Richard, M.G. (1984) "The Use of a "Selector" for Bulking Control at the Hamilton, Ohio, USA, Water Pollution Control Facility," *Water Sci. Technol., 16 (Vienna),* 35.

White, M.J.D. (1975) "Settling of Activated Sludge," Tech. Rept TR11, Water Res. Centre, Stevenage, England.

White, M.J.D. (1976) "Design and Control of Secondary Settlement Tanks," *Water Polln. Control, 75,* 459.

White, M.J.D., Tomlinson, E.J. and Chambers B. (1980) "The Effect of Plant Configuration on Sludge Bulking," *Prog. Water Technol., 12:3,* 183.

Williams, T.M. and Unz, R.F. (1983) "Environmental Distribution of *Zoogloea* Strains," *Water Res., 17,* 779.

Willis, B.A. (1988) "Mineral Processing Technology," 4th edn, Pergamon Press, Oxford, England.

Wilson, T.E. (1983) "Application of the ISV Test to the Operation of Activated Sludge Plants," *Water Res., 17*, 707.

Wilson, T.E., Ambrose, W.A., and Buhr, H.O. (1984) "Operating Experiences at Low Solids Residence Time," *Water Sci. Technol., (Vienna), 16*, 661.

Wood, D.K., and Tchobanoglous, G.T. (1974) "Trace Elements in Biological Waste Treatment with Specific Reference to the Activated Sludge Process," *Proc. 29th Ind. Waste Conf.,* *Purdue Univ.* West Lafayette, IN., Eng. Extension Ser., *145*, 648.

Wu, Y.C., Hsieh, H.N., Carey, D.F. and Ou, K.C. (1983) "Control of Activated Sludge Bulking," *J. Env. Eng. Div. Amer. Soc. Civil Eng., 110*, 472.

Zeng, D.R. and Pivnicka, J.R. (1969) "Effective Phosphorus Removal by the Addition of Alum to the Activated Sludge Process," *Proc 24th Ind. Waste Conf., Purdue Univ.*, West Lafayette, IN., *24*, 273.

Index